Traditions of Science Mediatization in Russia in a Global Context

Traditions of Science Mediatization in Russia in a Global Context

By

Yuliya B. Balashova

Cambridge
Scholars
Publishing

Traditions of Science Mediatization in Russia in a Global Context

By Yuliya B. Balashova

This book first published 2019

Cambridge Scholars Publishing

Lady Stephenson Library, Newcastle upon Tyne, NE6 2PA, UK

British Library Cataloguing in Publication Data
A catalogue record for this book is available from the British Library

ISBN (10): 1-5275-3029-9
ISBN (13): 978-1-5275-3029-4

Dedicated to my parents, who celebrate their diamond anniversary together

TABLE OF CONTENTS

ACKNOWLEDGEMENTS

First of all, I deeply share the ideas of Russian philosophy regarding the openness and fullness of the Russian soul, its mission as the unifier of mankind.

I believe that I was lucky, because life forced me to go beyond the basic philological specialty. However, philology turned out to be an excellent basis for the transition to a broader interdisciplinary sphere. In this book, I try to combine social science and the humanities. As an internal environment that has shaped me, I must single out my native home, Saint Petersburg State University. At the same time, and this is really important, I happened to get a good international experience.

The author gratefully appreciates the support of the Russian Foundation for Basic Research (RFBR); project No.16-03-50128.

INTRODUCTION

At the present time, an interest in the scientific problem has increased in the Russian media. Scientific subjects are being actively developed in the printed press and electronic media, and new popular science journals have been created. For the first time in a long time, the media in Russia has taken an interest in scientific issues; but modern science is capable of successfully developing only in a global context. The globalization of scientific knowledge contributes to awareness not only of the common threats but also of the desire to bring science closer to the interests of society. But along with this global trend, "science with and for society," a persistent negative phenomenon, such as the rejection of science, or a resistance to science, has clearly manifested itself.

> Yet the science of our day is more powerful than ever before. Computers are good for criminals and hackers as well as scientists and students. Science and technology can be abused as easily as they can be used for our common good. We need good scientists, but we also need good citizens who will ensure that our science will make the world a better place for us all to live in (Bynum 2013, p. 256).

In the conditions of the traditional reduction of constructive social and political discourses in Russian society, modern Russia is gradually entering (as the "creative class" positions it) an "era of enlightenment" and "the time of art". Throughout the Russian cultural existence, the literary model of the world replaced the social, while today popular science has acquired the qualities of such a constituent element. However, it must be borne in mind that Russian history is cyclical, and the revival of foundational elements of national identity, to which science belongs in its social expression, inevitably revises time-tested models. In general, the idea of cyclic development is one of the most important for 20th century science; for example, the economic cycles of Nikolay Kondratiev (K-cycles), or the socio-cultural dynamics in the interpretation of Pitirim Sorokin.

For the first time in the post-Soviet period, the state government has started supporting and initiating projects related to science popularization. This is extremely important because, unlike in developed Western countries, traditionally in Russia the state, and those affiliated with state institutions, play the primary role in scientific development. As a consequence, the media are beginning to pay more attention to scientific education.

Knowledge at all times is a value. The problem of science popularization exists as much as the problems of science itself. Historically, considerable experience of promoting science has been accumulated in Russia, a country with a rich history. Traditionally, science held a high position in Russian society and has been included in the public sphere. But, at the same time, it is important to take into account the differences between Russian and Western traditions and mindsets. The traditional forms of science mediatization include scientific societies, museums, libraries, educational films and lectures, the sphere of non-formal education, and general enlightenment. These communicative channels are extremely significant for the performance of science, especially if the agenda is the dissemination of knowledge about reality and the struggle against pseudoscience. Popular science magazines in Russia should, therefore, focus on the integration of science, and the synthesis of science and human knowledge. In order to master scientific knowledge, a mass audience requires a mediator. In the science communication process, the communicator is even more important than in politics, as politicians themselves must be able to act in the public sphere, including their contact with the press. In terms of communication theory, it is difficult to find any other field of public life where the communicant plays such a significant role. The reduction of communication with the public can have catastrophic implications for science itself as well as for society.

> Because government scientists essentially work for the American public, government communicators have an obligation to engage proactively with the media and encourage scientists to talk to reporters. Scientists are often thrust into the public spotlight unintentionally and may be unprepared and unsure of how to respond to media inquiries. It is our responsibility to help them learn how best to interact with reporters and tell their stories of discovery in plain language (Blum, Knudson and Henig 2006, p. 282).

To date, there is a clear global trend to distrust the traditional media. However, in Russia the bulk of the population still follow the custom of significant media consumption, something formed back in Soviet times. Accordingly, the new creative class that has appeared in Russia receives information mainly from the Internet. The new Russian system of science popularization is, thus, oriented towards this creative class. Moreover, Russian people traditionally have a huge latent interest in science, which is confirmed by Russian history. Therefore, traditional media such as socio-political television and newspapers are also trying to pay more attention to popular science issues.

There is a general transition in populist science education to more fully understand and utilize new media, which often turn out to be the "forgotten old". This trans-media or combined media can be divided into two conditional groups: various network resources, where absolute natural science subjects dominate; and cultural/leisure institutions, where the voice of humanitarian enlightenment and the game element are acquired. The infrastructure of cultural and educational clusters, intellectual clubs, salons, and cafes is expanding. In the development of these forms, there is undoubtedly a connection, both with Western practice and the elitist "pan aestheticism" of Art Nouveau. At the same time, bypassing the middle of the 19th century—with the scientific societies, public lectures and readings, and flourishing scientific congresses—this practice refers to the initial stages of the formation of the European scientific enlightenment, marked by the dominance of oral and visual forms (scientific societies, public lectures, and museums). Popular science, located on the cultural and educational portals (such as the Russian *Arzamas*), in the format of intellectual battles (as popular worldwide *Science Slams*), or in various educational clusters (*Ohta Lab* in Saint Petersburg). In many ways, these projects continue the central idea, typical of the whole Russian science mediatization tradition, of a commonwealth of different sciences.

But we are not at all inclined to hide or conceal the problems. In fact, they are fully reflected in the discussions on popular science in the Russian public sphere. The public discussion of science in Russia did not go the way it went in the leading Anglo-Saxon countries. Therefore, in the case of this discussion's latest resumption, a kind of return to the starting point takes place. The crucial issue is the old Darwinism/creationism polemic, which acquires the meaning of opposing science and religion. Religion begins to be identified with pseudoscience, which avoids or rejects any discussion of this issue, at least in the Soviet era. In the modern West, beneath the ruling concepts of tolerance and multiculturalism, the science/religion dichotomy loses much of its force. For example, it is believed that religion does not contain anything that would deny scientific knowledge, and vice versa. At the same time, the famous science popularizer Richard Dawkins, instead of the "illusion of God", seems to be proposing something resembling a new cult of rational knowledge. This latter approach was adopted by a new generation of Russian scientific popularizers and provoked justified criticism. It should be also borne in mind that in Russia the aspiration of scientists to counter-pose science and religion is further conditioned by a protest against the growing clericalization of all aspects of society. At once, we should not forget that throughout the Soviet era it was science that replaced Soviet atheists with a repressed religious consciousness. Even earlier, at the turn of the 19th and 20th centuries, the real flowering of popular science was directly caused by the positivist worldview crisis. In the designated sense, the opposition of science and religion is especially destructive. It is known that the Dalai Lama repeatedly spoke at the Congress of the Society of Neurophysiologists, and in every way tried to establish a connection with the world scientific community (which is not accepted by all scientists).

Media discussion about public science in modern Russia is not simply unfolding on the model of network communication, but in a meaningful way it forms a hermeneutic circle. Based on the well-known position that the level of science communication directly reflects the level of public sphere development, the low level of public discussion about popular science at the moment inevitably corresponds to the unsettled relationship between science and society in Russia today. Thus, the condemnation of humanistic knowledge, absolutely impossible in a normally functioning society, becomes a key feature of the Russian popular science field. Public reputation, as the only real principle of a modern scientist, giving, among other things, a basis for the monetization of scientific activity, ceases to be a socially significant phenomenon. The other side of this negative phenomenon is exaggerated adherence to scientific ethics. Something that helps to explain Grigory Perelman's (the famous Russian mathematician who proved The Poincare Conjecture) fight against the global scientific bureaucracy. The very personality of the scientific popularizer (communicator) is problematic. Since in Russia at present this area has received some money, various fake scientists and journalists, who just pretend to be scientific activists, have rushed in. Hence, a natural question arises that is relevant to the Russian scientific community: is it possible for non-experts or journalists to popularize science?[1] The answer to this question is well known: yes, of course, it is absolutely possible and even recommended. However, this self-evident response has to be, first of all, convincing to Russian science itself. The modern global media world is usually described in terms of "post-truth", which is fully applicable to the sphere of public science. However, the most important role of science popularization is precisely in overcoming scientific castes and sectarianism. Science itself is elitist and gravitates toward isolation and away from amateurs. But in this case, the rupture of the celestial scientists

---

[1] See Balashova, Yu.B. 2017. Debate about Scientific Popularization in Russian Public Sphere (Based on Grey Literature Material). In: *The Grey Journal: An International Journal on Grey Literature,* vol. 18, n. 1, pp. 179–181.

and the ordinary people is only intensifying, and mankind is at risk of being left in a stone jungle. This situation is often presented in Sci-Fi movies, which gives a futurological forecast of the final degradation of most of the world's population.

If John Horgan's views on the "end of science" (Horgan 1996) sometimes menacingly resemble reality, then the enlightenment, the cultural break, begins to experience its revival as a specifically Russian project of salvation. In such a syncretistic country as Russia, science popularization invariably turns out to be a kind of broad enlightenment within the conditions of a non-modernized society. At the same time, science and science communication clearly demonstrate how global the world is.

References

Balashova, Yu.B. 2017. Debate about Scientific Popularization in Russian Public Sphere (Based on Grey Literature Material). In: *The Grey Journal: An International Journal on Grey Literature,* vol. 18, n. 1, pp. 179–181.

Blum, Deborah, Knudson, Mary, and Henig, Robin Marantz. 2006. *A Field Guide for Science Writers*. New York: Oxford University Press.

Bynum, William. 2013. *A Little History of Science*. New Haven and London: Yale University Press.

Horgan, John. 1996. *The End of Science: Facing the Limits of Knowledge in the Twilight of Scientific Age.* Reading, Mass.: Helix Books / Addison-Wesley.

PART I

THE MEDIATIZATION OF SCIENCE: HISTORICAL PHENOMENON AND SOCIAL PRACTICE

CHAPTER ONE

HISTORICAL AND THEORETICAL APPROACHES TO SCIENCE MEDIATIZATION INTERPRETATION (METHODOLOGY, CHRONOLOGY, FORMS OF SCIENCE MEDIATIZATION)

One of the greatest scientists and science popularizers in the contemporary world, Stephen Hawking (who passed recently and was buried in Westminster Abbey near the graves of Newton and Darwin) repeatedly said that in the modern world there was no single theory explaining everything. As is known, science originated in the philosophy of nature and its viewpoint. Throughout Antiquity and the Middle Ages, scientific knowledge existed as a complex, connecting the physical and anthropological aspects of the world. The craft of a scientist was conceived as universal, and this type of scientist was exemplified in encyclopedic figures, from Aristotle and Leonardo da Vinci to Lomonosov. Philosophical rationalism proposed strict logical boundaries between objects and phenomena; and later, within the conditions of the development of industrial society, the task was to create a comprehensive positivist system of sciences. Quite recently, in the second half of the 20th century, the universal scientific methodology of semiotics was developed and successfully applied, within the framework of which it was possible to consider different systems (social, humanitarian, and technical) from the unified position of a doctrine of signs. In 1980, the famous futurist Alvin Toffler in his "classic study of tomorrow" predicted:

> Today I believe we stand on the edge of a new age of synthesis. In all intellectual fields, from the hard sciences to sociology, psychology, and economics—especially economics—we are likely to see a return to large-scale thinking, to general theory, to the putting of the pieces back together again. For it is beginning to dawn on us that our obsessive emphasis on quantified detail without context, on progressively finer and finer measurement of smaller and smaller problems, leaves us knowing more and more about less and less. Our approach in what follows, therefore, will

> be to look for those streams of change that are shaking our lives, to reveal the underground connection among them <...> Our job here will be to think like generalists, not specialists (Toffler 1990, p. 130).

In the modern scientific environment, a constant talk about "new evolutionary synthesis" takes place:

> Many believe that the development of biology is constrained by the lack of an adequate theoretical basis, a comprehensive new theory that could make the search for new knowledge more meaningful and constructive (Markov 2015, pp. 18–19).

Richard Dawkins extrapolates biological processes to the dissemination of cultural information (dichotomy: "gene/meme"), and has defined science as the "magic of reality" (Dawkins 2011). These noted positions, based on the idea of knowledge convergence, are conceptually fundamental to the present study. The model of knowledge in the West is based on the separation of science/art. The ideologue of capitalist society, sociologist Max Weber, made a distinction between universal and narrowly specialized knowledge in the definitions: "science as a vocation and profession" (Weber 1946). In a logocentric and metaphysical country such as Russia, these dichotomies are not entirely justified. Science communication existed in Russia at the beginning of the 18th century (all European processes took place in Russia with a certain delay; a similar lag in general was overcome by the beginning of the 19th century) basically as popular science communication. Typically, the Russian national consciousness tends toward traditionalism and syncretism. In addition, Russia has not had the historical preconditions for the formation of narrow specialization. One of the most significant cultural reasons for this, as compared with Western Europe, is the less consistent development of a classic Russian hierarchy (Kondakov, Sokolov, and Hrenov 2011). However, at the same time, the fundamental scientific works of Western scholars in the humanities (such as Marshall McLuhan, Alvin Toffler, Francis Fukuyama, and others) are not recognized in Russia as actual scientific discourse, but, rather, as a form of essayistic scientific journalism.

In the West, Russia is considered to be a country with an archaic socio-political system, while at the same time having an avant-garde: advanced science and progressive art. In a country gravitating toward syncretism, such as Russia, science and art are closely connected. Science popularization in Russia has consistently turned out to be a form of general enlightenment, a cultural activity. This is the most fundamental difference from those countries that lead the way in science popularization in Anglo-Saxon countries. Thus, the definition of "science communication" in

Russia is wider than in the West, organically including a humanities component (while the definition itself around the globe is only applicable to different branches of science).[2] This is why in Russia such cultural forms as travel or adventure literature enter the semantic field of the SciCom concept. In 19th century Russia, popular science works were published on a regular basis first in encyclopedias and then in classic large-volume magazines. The idea of the commonwealth of sciences was central to the whole tradition of popular national science journalism. In the classic Russian 19th century "thick", encyclopedic magazines: *Sovremennik (Contemporary)*, and *Otechestvennyie zapiski (Fatherland Papers)*, the departments of politics, science and literature were mixed. The first issue of the most famous Russian popular science magazine was published in 1890. We refer to *Nauka i zhizn' (Science and Life)*, which positioned itself as a "literary, artistic, social and popular science magazine". All subject areas were represented as forms of knowledge and cognition. The subsequent rise of the magazine's popularity during the Soviet era was determined by the fact that its audience was formed from the Soviet intelligentsia, who wanted to learn how things were developing in other areas of knowledge. In this popular science magazine, science was presented as knowledge of the whole world. The approach to the understanding of scientific knowledge as universal knowledge was typical of classical Russian popular science magazines and the entire tradition of science mediatization.

So, the first methodological criterion of this study is the syncretism of Russian popular science. This thesis is an original projection of Karl Jaspers's ideas in *The Origin and Goal of History* (Jaspers 2010) on modern scientific universalism.

Another important thesis: Russian science has never been separated from public life. The Russian system of secondary and higher education was based on the German model of education, named the "Humboldt model".[3] However, unlike their Western European counterparts, Russian universities were never "states within states"; in other words, they were never separated from the public.

In respect of the interaction: science vs. society, the biography of Dmitry Mendeleev is representative. A recognized scientist, influencing

---

[2] A broad cultural approach to science communication is presented in the book: *Communicating Science: Professional, Popular, Literary* by Nicholas Russell.

[3] The German system was also made into the basis of the university scientific staff training in the United States. The transition to the independent system took place in the 1900–1930s. See *Nauka po-amerikanski: Ocherki istorii* [Science in America: Essays on History]. 2014. Moscow: NLO.

the minds of young people studying in the 1880s, Mendeleev also acted as a public person. He was the author of the scientific and journalistic book "To the Knowledge of Russia: Treasured Thoughts"; he was also an active member of various societies. In 1890, he retired from St. Petersburg Imperial University after trying to defend students' rights. The story was that Mendeleev agreed to transfer to the Minister of Education the students petition demanding university autonomy (which once again was canceled). The Minister refused to accept the petition, and in response, Mendeleev did not consider it possible to continue serving in the Ministry of Education. He left the university, despite the fact that the Council of the university turned to him with a request not to leave.

We could give another example. In Russia, even methodological schools maintained themselves through the use of journalistic discourses. The Russian formalist school laid the foundation of valid literary criticism in the 1920s, yet also tended toward various forms of publicity. It is also significant that in Russia a unique socio-cultural type, such as the "intelligentsia" was formed. The basic quality of the Russian "intelligentsia" is the realization of a moral obligation to society. The implementation of this moral debt resulted in widespread enlightenment (the organization of schools, educational societies, and so on). At the turn of the 19th—20th centuries, enlightenment initiatives targeted at various social strata became common. At that time, the main organizers of cultural and enlightenment activities were patrons who, having a different social status, established worker's enlightenment societies, people's universities, and people's houses. Many Russian intellectuals (professors, pedagogues, lawyers, doctors, etc.) took an active part in the different enlightenment activities, thus realizing the moral and ethical concept of "the 'intelligentsia's' duty to the people". A somewhat different approach to enlightenment and the popularization of science was established in the Soviet era (more precisely, in the second half of the 20th century), when outstanding Soviet scientists (not just enlighteners or the "intelligentsia") participated in the work of an extensive network of various scientific societies and associations.

Traditionally, Russian science was public and society-oriented; and such a special role still belongs to popular science journalism.

At the same time, it is necessary to take into account that traditionally in Russia the state has played and continues to play a huge role in the various spheres of social life. This is a fundamental fact that distinguishes the Russian system from the Anglo-Saxon model (where private initiative and business are prioritized). In Russia, without any support from the state, no major technological project could take place (such as *Skolkovo*). This is also true in the popular science sphere. So, the famous Russian travel magazine, *Vokrug Sveta* (*Around the World*), which has been published since 1861, is disseminated today by means of a mandatory subscription arranged with Russian schools. This decision was directly taken by President Vladimir Putin as an honorary trustee of the *Russian Geographical Society* which publishes the magazine. In this context, the fate of non-state projects is very problematic (such as, for example, Dmitry Zimin's *Dynasty Foundation* with a wide range of educational initiatives).

In terms of the development of civil society, science coverage represented an ideological niche. In certain epochs, it was precisely in popular science journalism that political polemics were concentrated (under the typical Russian conditions of the prevalent reduction of political discourse); in others (as in Soviet times), this sphere was less loaded with the inevitable propaganda. Traditionally, the scientific and popular science media were censored much more mildly than socio-political ones. During the Soviet era, the presenter of the famous television program *Ochevidnoe—Neveroyatnoe* (*The Obvious—The Unbelievable*) often allowed a polemical style of discussion. This was not typical of the dogmatic Soviet press as a whole. The Soviet press paid much attention to scientific life, but during the periods of liberalization (Khrushchev's "thaw", Gorbachev's "perestroika" (reconstruction)) the problems and shortcomings of scientific life were also discussed. Even in Soviet times, the Academy of Sciences often independently made decisions that contradicted Communist party directives.

The question of periodization also has a clearly expressed methodological nature. In this regard, in Soviet times, an ideological manuscript by Eleonora Lazarevich, *Populyarizatsiya nauki v Rossii* (*Science popularization in Russia*, 1981) was typical, while more recently, a journalistic book by science journalist Andrey Vaganov, *Spiral' zhanra: Ot "narodnoj nauk"' do razvlekatel'nogo biznesa: Istoriya i perspektivy populyarizacii nauki v Rossii* (*Spiral of the genre: From "folk science" to the entertainment business: History and prospects of the popularization of science in Russia*, 2014) demonstrates the changing trend. Andrey Vaganov's popular book is the first holistic attempt at a popular scientific historical construction undertaken in the post-Soviet period. The central idea of Vaganov's book is that "public interest in science depends little on the circulation of popular science literature. It is precisely industrial development that leads to the development of popular science periodicals and a literary system. Not vice versa" (Vaganov 2014, p. 151). In other words, science popularization is a direct consequence of the industrial development of society. In accordance with the stated proposition, the author builds a certain genre evolution and typology of the popular science genre "... as a concrete historical phenomenon. It passed several stages in its development and continues to change" (Vaganov 2014, p. 23). Vaganov suggests the following scheme of genre development, highlighting four main stages. The first stage is connected with the scientific revolution in Europe and covers the 17th—early 18th centuries. Vaganov calls this period: "Popular Science, but ... without the popularization of science proper" (Vaganov 2014, p. 23). The problem is that the given examples do not fully support the thesis put forth by the author. Thus, emblematic for the epoch, *Mesyachnyie istoricheskie, genealogicheskie i geograficheskie primechaniya v vedomostyah (Monthly historical, genealogical and geographical notes in the Gazette*) (1728—1742), published by the Academy of Sciences and Arts in St. Petersburg, is correctly characterized by the author as "... designed for a narrow circle of readers, first of all—for the academicians themselves, and not intended for the general public" (Vaganov 2014, p. 28). One gets the impression that if we use the conventional concept of "folk science" then it would be better to characterize various late Medieval texts, and those transitioning to the New Time period miscellanies, as "herbalists" or medical books. In the second period of the popular science genre development proposed by Vaganov, he devotes his attention to very specific time limits: 1750—1850. And as regards his basic definition of this period as "industrial enlightenment", caused by the industrial revolution in Europe, it is probably worth agreeing. But it's difficult to agree with the dominant "historical form of the popular science genre"—"booklets on technical amateurs support" (Vaganov 2014, p. 24). Although by no means a genre, the main type of publication at that time was the encyclopedic journal if we are to understand the textual channel for the dissemination of information within the category of science popularization. And here one more question arises: what are we talking about: a genre or a super-genre? Apparently, at first, historical types of science popularization (collections magazines, almanacs) were developed, and then (rather later, at the turn of the 19th—20th centuries) —a popular science genre as such. The third period of the proposed genre evolution is determined by Vaganov in a temporary relationship, from the mid-19th to the mid-20th centuries, and fits into the "entertaining science" formula. Such a description does indeed seem justified, as it was formed by positivistic ideas regarding the value of knowledge, this period—at least in Soviet Russia—lasted until the end of World War II, due to the powerful influence (personal, paradigmatic, and institutional) of pre-revolutionary science. Finally, the fourth period, according to Vaganov's classification, covers the middle of the 20th—beginning of the 21st centuries and is defined as "Popular Science 0.2", as part of the entertainment business.

Andrey Vaganov insists on "the bad genre genetics in Russia" (Vaganov 2014, p. 38). If we approach this genre interpretation from a technical point of view, then the Russian tradition really does not compare with German history, as proposed in the book. If, on the other hand, we include the very typical Russian humanistic component in our genre interpretation, then the picture turns out to be quite different. Thus, we need to link the history of science mediatization in Russia with the socio-cultural tradition and the phenomenon of "folk science". A truly mass audience, ready for scientific knowledge, appeared in Russia only in the 1930s. In the late 1930s, physics Nobel laureate Pyotr (Peter) Kapitsa wrote a letter to the "folk leader" Stalin. The genre of "writing to the tsar" is very characteristic of the Russian tradition, indicating the limited channels of communication with the authorities. From 1921 to 1934, Peter Kapitsa worked in Cambridge under the leadership of Ernest Rutherford. He was a full member of *The Royal Society of London for the Improvement of Natural Knowledge.* Kapitsa considered the system of science organization in England to be exemplary. So, in his letter to Stalin, Kapitsa directly related the level of scientific development to scientific propaganda, emphasizing that "the masses have a great natural interest" in science:

> Comrade Stalin,
> Our situation with science is adverse. All the usual public statements that claim that the Soviet Union offers the best conditions for science are untrue. These statements are not only bad because they are lies, but they are even worse because they do not allow us to get the scientific life in the country up and running <…> In my opinion, the goal is straightforward: we need to foster mass interest in science and show its significance for progress. I do not think this is too difficult since the masses have great natural interest <…> Capitalist countries pay much attention to scientific propaganda. Such work is especially widespread in England, which, in my opinion, largely explains <…> its exceptionally high level of science. England established special societies to popularize science—the Royal Institution and the British Science Association—one hundred years ago. Its museums—the British Museum and the Kensington Museum—are the biggest in the world; its press pays more attention to science and scientific life than any other country does… In the Soviet Union, the popularization of science is botched. England's example is quite illuminating <…> I am certain that if we manage to interest the masses in science, then the scientific workers will become enthusiastic. They will become the pride of the country, they will be proud of Soviet science, they will organize it themselves <…> (quote: Akopov, 2012, pp. 116–117).

In this letter, Peter Kapitsa also identified five main directions of scientific propaganda: scientific museums, movies, popular literature and lectures on scientific subjects, scientific journalism, and the propagation of science in schools. Conservatism, which is typical of a system for disseminating scientific knowledge that is intent on protecting the elitist essence of science, made it so that the main means of popularizing science had remained virtually unchanged since the beginning of the modern era. These include 1) the press; 2) scientific debates; demonstrations, public lectures, defenses of theses; 3) scientific societies (the least accessible form); 4) museums; 5) the educational system.

In Russia, as we have noted, the printed channel for the spread of scientific knowledge was very important. However, in a global context the oral tradition of knowledge dissemination, which is extremely relevant today, remained primary right up until the end of the 19th century. The most closed elite form is indeed the scientific community. The development of science communication is, however, closely connected with the growth and complexity of the public sphere. However, in Russia, due to the significant role of the state, we see a model of cooperation between civil society and the authorities. The American historian Joseph Bradley has written one of the most comprehensive books, one that also discusses the scientific societies in Tsarist Russia. In the editorial introduction to the book's translation into Russian, the following thesis is proposed:

> Considering the vital activity of societies in the sphere of science and education, Bradley calls for the consideration of them as a model of self-organization, within the framework of which a public sphere of civil society was formed and an appropriate political and legal culture, civil consent was attainable. <...> Joe Bradley considers the basis of civil society voluntary associations, sharing the opinion of the leading theoretician of the public sphere, J. Habermas. <...> Meanwhile, Bradley refutes the widespread view of the political culture of the Russian Empire as completely confrontational, based on the confrontation between a strong state and its weak and politically immature subjects <...> (Bradley 2012, pp. 6,7).

The golden age of museums was the 19th century when most of Britain's great museums were founded; e.g., "… museums <…> private and public <…> built for entertainment, for education, and for research.<…> amorphousness of the now-familiar categories in the 19th century and <…> the then-undetermined meaning of 'museum' <…> sites of miscellaneity" (Science Museums 2017, p. 10). The leadership in the creation of science museums really belongs to England, with the first science museum originating in Oxford in the late 17th century—*Museum of the History of Science*. In 1714, in St. Petersburg—the newly emerged capital of the Russian Empire, *Kunstkamera* was opened, a museum of rarities (now anthropology and ethnography). It was also the first public state museum in Russia. All developed countries in the contemporary world have scientific museums (London, Amsterdam, Chicago, etc.), and, one thing that defines them all: they must be interactive. For instance, the *Franklin Institute* in the US was founded in 1824 "for the promotion of the Mechanic Arts". On this base, known today to the whole globe, the Franklin Institute Museum opened during the Depression in 1934, and was called the "Wonderland of Science". A hands-on approach to science learning was established at that time. Today, at the Franklin Institute Museum in Philadelphia are presented "live-learning" programs and a large gallery for traveling exhibitions. Science museums are now expected to form a creative urban environment.

Traditionally, the main channel of science popularization (a position it still retains) was considered to be certain segments of the press, such as, for example, popular science magazines. Special editions of Journals also appeared in a scientific format. In 1665, in Europe, the first two journals were published within a few weeks of each other: first in Paris, the *Journal des Scavans* was printed; and then in London, *Philosophical Transactions: Giving Some Accompt of the Present Undertaking, Studies, and Labours of the Ingenious in Many Considerable Parts of the World* (Vol. I. For Anno 1665, and 1666), initiated by the *Royal Society of London for Improving Natural Knowledge*. The Royal Society of London is one of the world's oldest scientific organizations. Members reported their results in lectures and publications, establishing conventions that are still followed today in the communication of scientific knowledge. Specialist journals appeared in succession: from the ten publications that comprised the first issue of the London Journal, three were reprinted from the French *Journal of Scientists*. One popular format of knowledge exchange had been provided by the genre of letters, as well as polemics, and literary and critical writings. In 1668, the first Italian journal appeared in Rome and was conceived by its editor on the model of the French, and later served itself as a model for Italian literary magazines. *Il Giornale de'Letterati (Literary Magazine),* which housed works of a literary, linguistic, and philosophical nature, existed until 1679; whereas the French *Journal of Scientists* lasted until 1828, and the history of *Phil. Trans.* continues to this day. Pitirim Sorokin's student and a follower of Max Weber, the famous American sociologist Robert K. Merton, established the origin of modern science in

England by rooting it in the dominance of the Protestant work ethic (Merton 1973).

The appearance of three important European magazines at the same time is due to a change in forms of social life and the nature of the philosophy of science in modern history. A leading position in the process of knowledge production began to be assigned to the scientific environment, in contrast to an orientation toward the sacred personality of the Teacher and timeless knowledge in the Medieval period. Nevertheless, it was medieval proto-science that became the foundation of modern science; and it is well known that the search for "the philosopher's stone" contributed to the development of chemistry. In this study, we are not interested in the history of science as such, but in the mechanisms of its interaction with society. In addition to the main oral channels of "popular science" spread during the middle ages, there were also handwritten analogs. Indeed, the most common channel for the spread of popular science over the centuries has been certain types of miscellany, such as, for instance, the almanac.

The transitional form of the almanac (located on the boundary between periodical and non-periodical publications) operates in different ways in different national traditions. However, regardless of the type of culture and civilization, the almanacs, due to cultural marginality, clearly preserved the closest correlation with their historical ancestor: almanacs originated in the Middle Ages in the East.

> Almanacs are one of the most ancient surviving forms of literature in the Western world. One of the earliest surviving copies dates from the period of Ramses II (1304—1168 B.C.E.) on papyrus held at the British Library. The origin of the word itself appears to be much later, possibly linked to the Arabic word for calendar <…> (Curth 2005, p. 258).

Almanacs appeared in Western Europe as a result of the Crusades. The very Arabic word *al-mana* means "time", "measure", and "calendar". The calendar, as the oldest instrument for physical time measuring, is the core of any cultural tradition (along with language, customs, and rituals). The inviolability of chronology is designed to ensure the stability of the world order: the ancient Egyptian pharaohs, when entering the throne, promised that nothing would change in the calendar. An almanac-calendar of the 14th—15th centuries was an astronomical ephemeris or a calendar table. It contained not only a list of days of the year, including holidays and weekdays, individual components of The Priests but also information about astronomical phenomena: the rising and setting of the moon and the sun, solar eclipses, the time of warm and cold weather, etc. To such forms of information, which were mostly household and agricultural recommendations, favorable days, and also astrological predictions were usually added.[4]

The medieval craving for astrology is due to the attitude to eschatology in the Middle Ages because calendars helped to orient the people regarding prophecies concerning the end of the world. The first printed almanac was compiled by the Austrian astronomer and mathematician *Georg von Peuerbach*, and was called *Pro annis pluribus (For Many Years)*.[5] For a long time almanacs developed into a platform for the convergence of folk science, mythological views, and church dogmas (see, for example, the famous astrologer William Lilly's[6] almanac of 1647, named *Christian Astrology*). However, the special popularity of the astrological predictions contained within such almanacs eventually provoked defensive actions on the part of the state and the church in both Western Europe and Russia, which contributed to their gradual (from the second half of the 16th century) repression. Thus, in France in the 1560s, the government issued a series of prohibitive decrees on almanac-calendars. However, the sacramental almanacs continued to enjoy influence even during the transition to modern times.

From the standpoint of modern concepts, the designated content of almanacs refers to pseudoscience, but here there were the roots of general scientific (astronomical, medical, historical, and philological) views. In a universal, encyclopedic type of proto-publishing—the almanac-calendar—a kind of universal proto-science was presented. "The almanac-makers of early modern England boldly took as there subject the whole of Creation", with this phrase the fundamental book on English almanacs by Bernard Capp begins (Capp 1979, p. 13). This book includes special sections, devoted to the almanacs' "Astrology, Science and Medicine", "History and Literature":

> Almanacs were cheap, popular annuals, and so by their very nature unsuited for an extended investigation of scientific problems. They contained, however, far more than the mere repetition of traditional ideas which might be expected. Many compilations showed a marked awareness

---

[4] About almanacs' general history see Almanac. In: *Encyclopædia Britannica*: in 24 vols. 1964. Chicago: Encyclopædia Britannica, Inc. Vol. 1, pp. 655–656; Almanac. In: *The Encyclopedia Americana*: in 36 vols. 1946. New York; Chicago: Americana Corp. Vol. 1, pp. 430–431.

[5] Magic symbolism is strong in modern almanacs as well. See *The Witches' Almanac.* 1974—1978. New York: Publishers Grosset & Dunlap.

[6] William Lilly lived in 17th century England and published annual almanacs with predictions.

> of scientific advances and were active in reporting them. The almanacs had an important role, especially in the seventeenth century, in the popularization of the new science. <…> They could play a valuable role, as in astronomy, in publicizing innovations. <…> one of the most common features of the Stuart almanac was the 'chronology' or the brief history of the world. <…> much of the humor was, in fact, less than savory in character and was based on racial, religious and sexual prejudices. <…> the astrologers' limited interested in literature and the other arts reflects partly the fact that the majority were self-taught men. <…> mediaeval astrology developed with little regard to national boundaries, and the history of almanacs too has international ramifications (Capp 1979, pp. 180, 200, 215, 234, 235, 270).

The traditional almanac's medical content was analyzed by Louise Hill Curth—

> …"the golden age" of English almanacs was between 1640 and 1700, because of their involvement with "political, social and religious controversies". These parameters also work well for the study of medical history, because this period begins with a burgeoning of medical and astrological books, made possible by the abolition of government censorship in a country moving ever closer to civil war. <…> This article is based on 1,673 almanacs that are known to have survived between 1640 and 1700, of which 1,392 (83.2 percent), have been examined. Almost three-quarters of these almanacs contain either preventive or remedial medical advice and/or advertisements for medical products or services. In general, this material focused on "popular" medicine, or traditional Galenic principles and practices, with barely a nod to the great scientific discoveries of the seventeenth century, such as circulation of blood within the body (Curth 2005, pp. 257, 258).

At the beginning of the 19th century, following France and Germany, a literary variety of almanacs was established in Russia as the dominant form (accordingly—*Almanach des Muses, Musen-Almanach, The Calendar of Muses*). In a number of European countries and also in Russia (which culturally, of course, was a part of European civilization), almanacs had become a communication channel for the promotion of national literature. In England, the historical development of almanacs was associated with their specialization (for instance, almanacs of the various societies, including scientific, etc.).

Let us give an example of the typical English almanac of the mid-19th century, which does not refer to a specialized, but a general type. In the preface it was stated:

> The present volume, though complete in itself, is the second of a series of the quarterly volumes, entitled the Chronicles of the Seasons, or the Progress of the Year, designed to furnish a seasonal account of the natural history and phenomena of each month, in conformity with the present state of knowledge, varied with notices of the arts, antiquities, and customs of our native country, and arranged in convenient portions for daily reading (Chronicles of The Seasons, p. 3).

Colonists brought almanacs with them to America (on American soil, a tendency towards politicization immediately manifested itself).[7] Today the U.S. Farmer's almanacs, or folk calendars, which have become an important part of popular culture, were very common. In the American tradition, it is understood that almanacs are a reference work for farmers; giving weather predictions, tracking the phases of the moon and suggesting optimal dates for planting various crops. Agricultural information and calendars were certainly central to this genre, but almanacs were also used to promote temperance, argue against slavery, list political officeholders or member churches in a denomination, and might include maps, health information, recipes, astronomical data, fiction, advertisements, and a variety of other material. Farmer's almanacs (which actually provided considerable content beyond the agricultural), often included significant health information, plus advertisements for patent cures. One of the best American almanac collections is situated at Michigan State University Library. Head of the MSU Library Special Collections, Patrick Olson said, "The collection of almanacs got immediate use by a class studying early astronomy. <...> People mostly tossed these almanacs away. I wanted the collection to take popular culture and push it back in time" (Lansing City Pulse 2018).

Modern forms of science communication, as well as public communication in general, gravitate toward dialogism:

> Examples of twenty-first-century dialogue events include *café scientifique*, scenario workshops, deliberative opinion polls, citizen juries, people's panels and in the US, consensus conferences. On the surface, these seem to be good examples of dialogue events, but closer scrutiny has revealed some problems with a dialogue-focused approach. In 2009, Sarah Davies examined informal public dialogue events at the Dana Center in London (a purpose-built center, and part of the Science Museum). These were panel events, where expert panel members spoke and then the public audience were able to comment and ask questions. What she discovered was that this format of comments, questions, and responses, was not a simple dialogue event, instead, these panel events had elements of both deficit

[7] "Much colonial publishing involved pamphlets and other occasional pieces" (The Media in America 1996, p. 46).

> and dialogue. This research indicates that a pure dialogue event is often difficult to achieve. In addition, it isn't clear how these examples of informal dialogue actually feed into government policy (Bowater and Yeoman 2013, p. 16).

Thus, we summarize the first chapter in terms of the concretization of the research methods utilized. As specific research methods, cultural-historical and comparative historical methods were used. In the presentation of the historical material, certain points stand out. We attempt here to work within the area of interdisciplinary research, connecting social and humanistic studies. Without going into the extensive theory of social mediatization, we note that we have chosen this term, bearing in mind the factor of transforming social and cultural life through the media. We believe that this media effect is the most desirable for science communication. Due to the variety and large coverage of material relating to fundamentally different media systems, we adhere to a unitary media-based logic (Couldry 2008). In this sense, the main focus is on the mediatization channels. Depending on the nature of communication between science and society, several models are distinguished:

> The linear model is a one-dimensional version of the two-dimensional diffusion model of popularization, in which the scientific community emits information to be received by the public. Communication scholar Robert Logan argues that the diffusion model is essentially a model in which science communication is intended to be purely persuasive (one might say propagandist); his study of health communication concluded that such communication does little for the public's health literacy. The diffusion model, like the dominant linear model, describes what sociologist Brian Wynne has called "top-down" communication, in which information that is meaningful in the scientific context in which it originated is likely to arrive in an entirely different lay or public context, in which people are unlikely to accommodate it (should they choose to) without substantial interpretation and adjustment.
>
> While linear and diffusion models have some explanatory power on a gross level, they contribute little to an understanding of the complex communication systems in which scientists routinely engage. For example, locating the "scientific" and the "popular" at opposite ends of a scale is different to maintain in practice (Gregory and Miller 1998, p. 87).

In any case, the effectiveness of the science mediatization process is mainly measured by the quantitative and qualitative development of in-demand media forms. The more diverse the forms of mediatization—the more obvious the media effects: this is what determines our approach. At the same time, we fully share an established position connected with the recognition of the media's growing influence on society and culture, which is commonly called "medialization" (Matheson 2005).

An appeal to the historically established patterns of interaction between science and society is extremely important also because science itself and, as a consequence, the sphere of its mediation is rather conservative. So, the existing historical practices are potentially reproducible. But, to date, there are no fundamental studies on the history of science mediatization, not only in the Russian but also in the English-speaking scientific community. In contrast to this, there are a large number of different science stories:

> In 1837, at the annual meeting of the Royal Society of London, William Whewell finally received the recognition he had been working hard to achieve for years. <...> He had just completed his book, *The History of Industrial Sciences*, which would go on to become highly influential in the shaping of science as a profession so beneficial to humanity that it should receive financial support from the governments (Cooper 2016, p. 3).

The existing works deal mainly with certain aspects (national, problematic, and chronological) of the history of popular science (see, for example, Mussell 2007, Daum 2009). Attention is drawn to the fact that Russia is completely out of the sphere of attention when researchers investigate science communication around the world (see Schiele et al. 2012, Bucchi 1998). In this sense, Russia, with its rich history and potential, is paradoxical outside the western public field, being predominantly represented in the West within the framework of confrontational political models. This study is the first research, both in Russia and abroad, to explore the Russian science mediatization tradition within an international context.

References

Almanac. In: *Encyclopædia Britannica*: in 24 vols. 1964. Chicago: Encyclopædia Britannica, Inc. Vol. 1, pp. 655–656.

Almanac. In: *The Encyclopedia Americana*: in 36 vols. 1946. New York; Chicago: Americana Corp. Vol. 1, pp. 430–431.

Apokov, A.I. 2012. Popularization of science as a crucial link in the propaganda and advancement of research. In: A.I. Apokov *Journalism: Research Papers, Articles, Notes, and Essays: Collection of Papers*. Rostov-on-Don: North Caucasus Higher Education Research Center of the South Federal University Academy of Pedagogical and Social Sciences, pp. 112–118.

Berkowitz, Carin, & Lightman, Bernard (eds.). 2017. *Science Museums in Transition: Cultures of Display in 19th-Century Britain and America*. Pittsburgh: University of Pittsburgh Press.

Bowater, Laura, and Yeoman, Kay. 2013. *Science Communication: A Practical Guide for Scientists*. Hoboken, NJ: Wiley-Blackwell.

Bradley, Joseph. 2012. *Voluntary Associations in Tsarist Russia: Science, Patriotism and Civil Society*. Moscow: Novyj hronograf.

Bucchi, Massimiano. 1998. *Science and the Media: Alternative Routes to Scientific Communications*. Abingdon: Routledge (Taylor & Francis Group).

Capp, Bernard. 1979. *English Almanacs, 1500–1800: Astrology and the Popular Press*. Ithaca, New York: Cornell University Press.

Chronicles or the the seasons, or, the Progress of the year; being a course of Daily instructions and amusement, selected from the popular details of the national history, science, art, antiquities, and biography of our fatherland; in four books. Book the second… Containing the months of April, May, and June. 1844. London: John W. Parker, West Strand. M. DCCC. XLIV.

Cooper, Caren. 2016. *Citizen Science: How Ordinary People Are Changing the Face of Discovery*. New York: The Overlook Press.

Couldry, Nick. 2008. Mediatization or mediation? Alternative understandings of the emergent space of digital storytelling. In: *New Media & Society*, 10 (373). Accessed July 2018. http://journals.sagepub.com/doi/abs/10.1177/1461444808089414

Curth Hill, Louise. 2005. The Medical Content of English Almanacs: 1640—1700. In: *Journal of the History of Medicine and Allied Sciences*, 60 (3), pp. 255–266.

Dawkins, Richard. 2011. *The Magic of Reality: How We Know What's Really True*. New York; London; Toronto; Sydney; New Delhi: Free Press.

Daum, Andreas W. 2009. Varieties of Popular Science and the Transformations of Public Knowledge: Some Historical Reflections. In: *A Journal of the History of Science Society*, 100 (2), pp. 319–332.

Gregory, Jane, and Miller, Steve. 1998. *Science in Public. Communication, Culture, and Credibility*. Cambridge, MA: Basic Books.

Jaspers, Karl. 2010. *The Origin and Goal of History*. Abingdon: Routledge.

Kondakov, I.V., Sokolov, K.B., & Hrenov, N.A. 2011. *Tsivilizatsionnaya identichnost v perehodnuyu epohu: kulturologicheskiy, sotsiologicheskiy i iskusstvovedcheskiy aspektyi* [Civilization identity in the transitional

era: cultural, sociological and art criticism aspects]. Moscow: Progress-Tradition.
Lazarevich, E.A. 1981. *Populyarizatsiya nauki v Rossii* [Science popularization in Russia]. Moscow: Moscow University Publishing.
Markov, A.V. 2015. *Rozhdenie slozhnosti. Evolyutsionnaya biologiya segodnya: neozhidannyie otkryitiya i novyie voprosyi* [The birth of complexity. Evolutionary biology today: unexpected discoveries and new questions]. Moscow: Corpus.
Matheson, Donald. 2005. *Media Discourses: Analysing Media Texts*. Maidenhead, Berkshire: Open University Press.
Merton, Robert K. 1973. *The Sociology of Science: Theoretical and Empirical Investigations*. Chicago: University of Chicago Press.
Mussell, James. 2007. 19th-Century Popular Science Magazines, Narrative, and the Problem of Historical Materiality. In: *Journalism Studies*, 8 (4). pp. 656–666.
Nauka po-amerikanski: Ocherki istorii [Science in America: Essays on History]. 2014. Moscow: NLO.
Putting the “special” in Special Collections. 2018. *Lansing City Pulse*. Feb. 22. Accessed March 2018. http://lansingcitypulse.com/article-15799-Putting-the-&lsquospecial&rsquo-in-Special-Collections.html
Russell, Nicholas. 2010. *Communicating Science: Professional, Popular, Literary*. New York: Cambridge University Press.
Schiele, Bernard, Claessens, Michael, & Shi, Shunke (eds.). 2012. *Science Communication in the World: Practices, Theories and Trends*. New York; London: Springer.
Sloan, David Wm., Startt, James D. (eds.). 1996. *The Media in America: a history*. Northport: Vision Press.
Toffler, Alvin. 1990. *The Third Wave*. New York: Bantam Books.
Vaganov, Andrey. 2014. *Spiral' zhanra: Ot “narodnoj nauk”' do razvlekatel'nogo biznesa: Istoriya i perspektivy populyarizacii nauki v Rossii* [Spiral of the genre: From “folk science” to entertaining business: History and prospects of popularization of science in Russia]. Moscow: Lenand.
Weber, Max. 1946. Science as a Vocation. In: Max Weber. *Essays in Sociology*. New York: Oxford University Press, pp. 129–156.

CHAPTER TWO

THE POPULAR SCIENCE PRESS IN RUSSIA AT THE TURN OF THE 19TH—20TH CENTURIES IN SOCIO-CULTURAL CONTEXT

The life-changing era of the early 20th century featured a scientific and technological breakthrough that affected not only the global landscape but also the daily lives of people. Albert Einstein's theory of relativity on the one hand and Sigmund Freud's psychoanalysis on the other transformed the understanding of both the physical world and the socio-psychological nature of the human being. The progress of civilization accelerated the pace of life rapidly. The capital of the Russian Empire became a true megalopolis: it was illuminated with electric lights according to the technique invented by electrical engineer Pavel Yablochkov (in competition with another Russian engineer—Alexander Lodygin). Yablochkov's "candles" were used to illuminate the streets of Paris; in London, they illuminated the Thames embankment. In the West, this invention was associated with Thomas Edison's name. It was this "great self-taught American" who managed to make incandescent lamps large-scale. The nobility in St. Petersburg moved to automobiles, while common people began riding trams. Telephones, telegraphs, and the radiotelegraphs (radios) of Alexander Popov (the first trial taking place in 1895 at the meeting of the Physics Department of the Russian Physics and Chemistry Society at St. Petersburg University) provided timely communication. In the West, Guglielmo Marconi is considered the inventor of the radio, since his transatlantic transmission turned out to be significantly more commercialized, despite the fact that Popov's and Marconi's wireless transmission experiments were conducted in parallel, and independently. At the turn of the 19th—20th centuries, the Russian Empire was one of the most developed world powers (including the economy), with a brilliant culture (which determined the fashion for everything Russian), and advanced science. Alongside the examples of Yablochkov, Edison, and Popov (Yablochkov, Lodygin, and Edison were even born in the same year—1847), Marconi proves that the prerequisites for breakthrough scientific discoveries had developed in different countries, and science had become sufficiently globalized. At the same time, the commercialization of science in the West even in this era was higher than in Russia. At the turn of 19th—20th centuries, a certain level of scientific communication was achieved, providing (along with the experimental culture of modernity) the possibility of scientific discoveries being introduced to the general public. So, for example, Einstein's theory of relativity was quickly perceived by the social environment, rather than the non-Euclidean geometry of Nikolay Lobachevsky less than a hundred years before.

At the turn of the 19th—20th centuries, the whole world was seized with a fever of discovery related to the North and South Poles, with the aspiration to master new geographical spaces in the extreme North and South. All polar expeditions of Russian travelers began in St. Petersburg. The St. Petersburg press paid much attention to polar expeditions, publishing portraits of researchers, reports on their fate and their discoveries. Interest in the bold gains of the Arctic, as well as the romance of science itself, was very high at the beginning of the 20th century. Even today, Arctic projects still have a high priority.

The "Modern" person began to feel his or her own belonging not only to the national culture but to all of humanity. As a response to the horrors of World War I, a desire was formed for the unification of mankind, and a world capable of acting together.

In the late 19th and early 20th centuries, the Russian Empire was a large colonial country; however, the level of education among the population lagged significantly behind that of the developed European states. Therefore, Russia launched a large-scale campaign to popularize science, which resulted in Russian experts being considered some of the best in the world during the Soviet period. The effect that science had on public consciousness was huge. The specific flair of the modern era, notable for its "pan aestheticism" and God-seeking, formed a special approach to scientific knowledge. The real heyday of such scientific and educational initiatives at the turn of the 19th and 20th centuries was determined by the natural sciences, a rational picture of the world crisis and the transition to a religious-mystical worldview, as was expressed by the outstanding Russian religious philosopher Nikolay Berdyaev.

> Faith in the God of science is now shaken. The crisis is not only at the top of philosophical thought but also at the bottom of positive scientific work. Not only philosophy and epistemology but science itself, from below, overcomes many idols and gods and shatters those generally accepted foundations of knowledge that naively fed the materialistic and positivist philosophy. The scientific crisis has gone so far that present-day science denies the very existence of matter, degrades matter in an area that it

> considered its own. It is worth mentioning the electromagnetic theory of modern physics, energy, etc. Modern science also shakes the foundations of Darwinism and the recently dominant theory of development. Vitalism more and more wins the mechanism. Overcomes the complacency of scientists, their dogmatic attitude to the unshakable foundations of science is replaced by a critical attitude. In the field of scientific knowledge, new phenomena are invading, which the official dogmatism of scientists has recently rejected as supernatural, allowed only by superstition (Berdyaev 2005, p. 43).

In his memoir, the greatest Russian symbolist writer Andrey Beliy[8] describes in detail the problem of the "scissors" that arose before him during his studies at Moscow University:

> From the second course, my position as a student facing the problem of specialization became very inconvenient; specialty, built by me, entitled "Methodology of Natural Science", did not exist; there was no spike of natural science with philosophy, natural in Germany, where the "doctor of natural science" was called "doctor of philosophy"; and we had: if you philosophize, study philology <…>; if you read Darwin, then keep quiet about philosophy. <...> I set myself the goal of mastering the method of comprehending facts in the spirit of a worldview built on two columns: one was aesthetics, the other was natural science (Beliy 1989, pp. 421, 381).

The influence of the new, non-positivist science was enormous. Members of various intellectual elites sought to provide scientific reasoning for the spiritual domain and esthetic activities. Such was the anthroposophical teaching of Rudolf Steiner, that he termed "spiritual science", and which found resonance in both Europe and Russia. If we rely on the "type of scientificity" concept used by Michel Foucault, then undoubtedly this typified transformation in the era of modernity (Foucault 2004).

In journalism, this trend was manifested in the emergence of such magazines as *Vestnik okkultnyih nauk (The Occult Science Bulletin),* 1907, which considered themselves to be popular science publications. Back in 1875, at St. Petersburg University, a "mediumship commission" was set up under the leadership of Dmitry Mendeleyev (1834—1907). This commission tried to find a scientific explanation for such a phenomenon as spiritualism, which arose as a specific mystical trend in the middle of the 19th century in the United States. The practice of rituals aimed at

[8] Andrey Beliy is also considered the founder of modern poetry investigation; he created a scientific circle to study the phenomenon of rhythm.

communicating with spirits spread rapidly throughout Europe and was popular among the Russian aristocracy (Emperor Alexander II himself was fond of it). Another outstanding Russian chemist, Alexander Butlerov (1828—1886), was interested in spiritualism too. Being the founder of the Psychoneurological Institute in St. Petersburg, Vladimir Bekhterev (1857—1927), studied paranormal phenomena. He not only created a scientific institute and a school in science, but his descendants (granddaughter Natalia Bekhtereva, great-grandson Svyatoslav Medvedev)[9] also became authoritative scientists who continued the study of the brain. According to a fairly reliable version, Bekhterev was poisoned in 1927, after a medical examination of Stalin and a diagnosis of "severe paranoia" (this version appeared immediately after Bekhterev's sudden death, and formed the basis of Lion Feuchtwanger's story, "The Story of the Physiologist Doctor B.").

Science was interpreted as the *Mir Bozhiy (World of God)* (1892—1906), as one of the most prominent Russian literary and popular-science magazines of the early 20th century was named. This magazine, like a number of similar ones, was intended for self-education. Thus, at the turn of the 19th and 20th centuries, the scientific sphere was expanding, including elements of occult knowledge, and was perceived not as a pseudoscientific so much as a controversial, para-science. As a result, leading scientists were actively involved in the public discussion of this issue. As a consequence, there was a rapid increase in communication between science and society, and this process was global.

This scientific and technological progress—urbanism—and the new concept of the human being, who came from the world of nature but lived in "concrete jungles," became the worldview and foundation for avant-garde "future art" movements (futurism, cubism, abstractionism, and, later, constructivism) that were important for Russia. Outstanding cultural

---

[9] At a joint conference of brain researchers and Buddhist monks Medvedev said the following, "In fact, there are two traditions of Western science—before and after Pythagoras. In the days of ancient Egypt, Pythagorean theory had not been proved, but the builders of the pyramids had known what the square of the hypotenuse was. Otherwise the structure would fall apart. Western science has gone the Greek way, where each provision requires proof. But this is not necessarily the way; it is often just powerful logic and sound knowledge that is required to build a certain hypothesis. That is how science is constructed in Buddhism and, incidentally, in this way Russian science has succeeded. Our predecessors I.M. Sechenov, I.P. Pavlov, V.M. Bekhterev built models, theories, which were confirmed experimentally only decades later. For example, the law of conservation of energy—the foundation of our model of the world - had not yet been proven." (Kokurina 2018)

figures took an active part in the preparation of popular science periodicals. Thus, the "literary, scientific and political" anti-war magazine *Letopis (Annals)* was published in Petrograd from December 1915 to December 1917. It was published with the close participation of Maxim Gorky (whose 150th anniversary was celebrated in 2018), and nourished throughout its life a special piety to science, apparently at a psychological level associated with the compensation for a lack of one's "own universities". At Gorky's request, the outstanding Russian naturalist, Kliment Timiryazev (1843—1920) became head of the magazine's science department. In addition to A. Blok, I. Bunin, V. Mayakovsky, H. G. Wells and other recognized writers, on the initiative of Timiryazev, physiological colleagues were invited to work on the journal. These included Ivan Pavlov[10] (1849—1936) and Ilya Mechnikov (1845—1916), who received the Nobel Prize in 1904 and 1908, respectively. Timiryazev in the popular science book "The Life of Plants" (similar to Alfred Brehm's "Life of Animals") saw the lack of existing scientific enlightenment as a consequence of dogmatism and the authoritarian approach to teaching the public. His enormous contribution to popular knowledge (as a popularizer, editor, and publicist) was a direct consequence of his democratic public position.

At the turn of the 19th and 20th centuries, enlightenment initiatives targeted at specific social strata became common. At that time, the main organizers of cultural and enlightenment activities were patrons of a

---

[10] "But the prize for the most bizarre use of Pavlov's findings must go to his compatriots in the Russian Army.

The standard task used by the Russians in the Second World War was no match for the German Panzer, designed by Ferdinand Porsche, of later sports-car fame, who would be arrested after the conflict for his use of slave labor in his factories. Nor were Soviet anti-tank weapons much use against the superior army of the Panzer so a new and terrifying weapon was invented. Thinking of Pavlov's result with dogs and food, the Russian tank regiments trained dogs to associate the underneath of tanks with food by starving them for a few days and then letting them out to see piles of meat and treats placed under stationary tanks with the engines running.

The idea was as Pavlovian as it was brutal; cement the tank-food association and then starve the dogs, strap mines to their backs and let them loose on advancing German army. But the Pavlovian link became too firmly rooted in the minds of the dogs so that they only associated the underneath of Russian tanks with food and, on their debut in 1942, packs of tooled-up dogs forced three brigades of the Russian army into a panicked retreat. It was later figured out that this was likely because the Russian tanks smelled of diesel whereas the German tanks ran on petrol, but by then it was all too late—the Russian Dogs of War were retired" (Graeme 2013, pp. 96–97).

different social status, who established workers' enlightenment societies, people's universities, and *Narodnyie doma (People's Houses). People's Houses* originated in England, but in a different form, i.e. as pubs for the poor people, called *The People's Palaces*. In Russia, People's Houses helped solve the educational problems and tripled in number thanks to state funding. However, there were also private charitable initiatives. For instance, the famous Sofia Panina People's House located in the deprived Ligovsky suburbs of St. Petersburg. The development of out-of-school establishments as independent institutions began in the early 20th century. For instance, the "Daytime Orphanage" and, later, the famous "Settlement" society were established at the premises of a special children's club in Moscow in 1905. The name of the society referred to the American practice of missionary settlements intended for enlightenment work. Many Russian intellectuals (professors, pedagogues, lawyers, doctors, etc.) took an active part in such enlightenment activities, thus realizing the moral and ethical concept of "the intelligentsia's duty to the people". A somewhat different approach to enlightenment and the popularization of science was established in the Soviet era (more precisely, in the second half of the 20th century), when outstanding Soviet scientists participated in the work of an extensive network of various scientific societies and associations. By investigating the popularization of science in the 19th and 20th centuries, it is possible to assess the cultural and educational level of the country's population during that period.

Ever since Peter I established the first scientific organization—the so-called *Petrovskaya Akademia (Peter Academy)*—Russian science was governed by the *Academy of Sciences*. In the early 20th century, the elite educational institutions became the leading scientific centers. This became possible due to the fact that scientific schools were established therein. In general, creating an environment that stimulates scientific research is the defining factor in the expansion of scientific knowledge. Progress in the development of scientific paradigms is closely related to the special atmosphere of universities. In Russia, "the idea of a university" (which was the title of classic works by John Henry Newman and Karl Jaspers on the phenomenology of universities) was associated with the issue regarding the autonomy of universities. Educational reforms conducted during the reign of Alexander I and Alexander II were followed by further steps taken in this direction in the early 20th century. After the revolution of 1905, Russian universities became autonomous, i.e. were granted self-government. University autonomy included the right to choose academic programs independently; positions at universities were elective (including that of the rector, despite that fact that all previous versions of university statutes had this position as appointive); universities had their own courts, typographies, and censorship. In addition, various types of cultural and leisure activities began to develop in the general education system in the early 20th century. For instance, "scientific and literary student clubs, choirs and orchestras, which used to be banned, were now considered advantageous <…> the academic life of students saw the introduction of corporate representation, which used to be considered seditious. <…> the first Russian revolution radically altered the organizational forms, practice, and ideology of student associations. The major role was given to faculty, inter-faculty, and city coalitions of student self-government, which were established upon the initiative of the students themselves and elected at student meetings based on party lists; these organizations protected the academic, economic, and civil interests of students" (Ivanov 1999, pp. 250–251).

Russian universities were closely related to the establishment and development of the Russian press. For instance, in the early 20th century, students published special scientific and journalistic digests. These are an under-researched layer of the early 20th century press. Students often attended lectures at other faculties and tried not to make their study purely utilitarian. In order to clarify this latter thesis, which contradicts the currently accepted view, it is necessary to cite a lecture delivered at a university by Russian chemist Dmitry Mendeleev:

> However, simply getting diplomas and getting acquainted with a subject is not why we are here and not why universities are established… there is another, higher aspect…which grants university knowledge a tone that should be called the spirit of the university… This spirit consists solely and entirely in one thing: the aspiration to learn the truth, by any means—not practical benefits, not personal improvement…—all these things are secondary, they are appendages… Understanding the truth in all its purity and perfection—this is the only spirit of the university". (Muravyeva 2004, p. 86)

In addition to his larger-than-life Renaissance nature, the personality of Mendeleev was attractive because he was an idol of the young students, a public figure, and a publicist. At the biggest higher educational institution in Russia—Saint Petersburg Imperial University—one could listen to the lectures of physicist Ivan Borgman (1849—1914), who became the first person to be elected Rector of Saint Petersburg University in 1905, chemist Lev Chugaev (1873—1922), zoologist Valentin Dogiel (1882—1955), linguist Jan Baudouin de Courtenay (1845—1929), and philologist Alexey Shakhmatov (1864—1920). Students of the historical and philological faculty were especially keen on the "Pushkin Seminar", held by Semyon Vengerov (1855—1920). A firm believer in the cultural-historical method, which was dominant at that time, Vengerov educated a series of brilliant cultural figures and philologists. The latter managed to establish a new methodological school: the formalist method in literary science. The seminar resembled a literary society without any formal order. In essence, it was a scientific discussion platform for the young that created the environment, the importance of which was mentioned above. Future formalists Yury Tynyanov (1894—1943) and Boris Eikhenbaum (1886—1959) took their views from the idea of "the gay science" (which was the title of one of the central books of Friedrich Nietzsche). S.A. Vengerov was not only a prominent Russian literary scholar but also a literary critic, publicist, and editor; he was among the first to implement accurate methods of analysis, after collecting an extensive catalog for his fundamental (but unfinished) "Critical Biographical Dictionary of Russian Writers and Scholars (from the foundation of Russian education to the present day)" (Vol. 1—6. SPb., 1889—1904).

The constructive changes of the early 20th century also concerned the secondary school. The organization of school affairs in the early 20th century shows that a simultaneous development of out-of-school education was required to immerse schoolchildren in the educational environment. Cultural and leisure activities that were popular in the early 20th century included sports, excursions, clubs, and the school press. This segment of the press was, in modern terms, a hybrid or inverse medium, where science was combined with journalism, enlightenment, and esthetics. Knowledge in this context is learned immanently, based on teamwork, imitation, and games (Balashova 2007). However, probably the biggest result that such enlightenment had on education was the fact that at the beginning of World War I the country came close to achieving a general education. When proceeding from enlightenment, which manifested itself through various social institutions, to a related subject, it is worth noting that the main channel of the popularization of science was, and still is, the press: more precisely, popular science magazines.

In 19th century Russia, popular science works were published on a regular basis, first in encyclopedic and then in the classical large-volume magazines. This trend persisted in the early 20th century, despite the fact that large-volume magazines lost their leading position. One of the best Russian pre-revolution magazines – *Russkoye bogatstvo (Russian Wealth)* (1876—1918)—was a literary, scientific, and social magazine. At that time, science was combined with criticism, for instance, in *Vesy (Libra)*, a symbolist scientific-literary and critical monthly magazine edited by V. Bryusov; or with journalism, for instance, in "Problems of Idealism"

(1902) and the "Vekhi: Collection of Articles on the Russian Intelligentsia" (1909), the biggest social and philosophical digests of that time. The prototypes of Russian popular science magazines included British illustrated magazines. They influenced the popular science press of Russia, since they were adapted by Russian illustrated weekly periodicals, considering the fashion for everything British that was prevalent at the turn of the 19th and 20th centuries, and was described brilliantly by V. Nabokov in his autobiographical novel *Conclusive Evidence*. Novelties of foreign science and literature were brought to the notice of educated modern readers on a regular basis. A symptomatic fact was the emergence of the *Science and Civilization News* segment in the popular small-volume illustrated magazine entitled *Vsemirnaya illyustratsiya (World Illustrated)* (1869—1898), which in combination with its supplements had a significant influence on the further development of popular-science magazines. In the early 20th century, newspapers started featuring the scientific society chronicler (reporter) position. Popular newspapers began using scientific agendas to form newsbreaks in terms of sensationalism (the circumstances in which a whale was caught, etc.). Mass periodicals generally gravitate towards the popular-science element. For instance, *AID-Info*, one of the first Russian tabloids launched during the "perestroika" period, still considers itself a popular science newspaper. This is additional evidence of the latent interest of the Russian general audience in science: the same interest mentioned by Pyotr Kapitsa. The Russian audience is traditionally interested in acquiring scientific knowledge, which gives ample opportunity for popularizing science, something that is under-utilized at present.

A significant segment of the popular scientific press of the early 20th century was formed by the publications of various scientific societies. These included a fairly authoritative weekly illustrated magazine published over a long period (from 1912 to 1937)—*Izvestiya russkogo obshchestva lyubitelej mirovedeniya (News of the Russian Society of Amateurs of the World)*. This magazine's extended history is typical of the conservative environment of the popular science press, although it is not completely typical of the short-lived popular science magazines of the early 20th century taken as a whole. The magazine continued to be published successfully in the Soviet period since one of the participants of the revolutionary movement was involved in its creation. Thanks to this revolutionary, in 1909 the extremely famous *Russkoe obshchestvo lyubitelej mirovedeniya (Russian Society of World Science Amateurs)* appeared or, more simply, the *Society for the Study of the World*, *World-studies*, or *Those Who Know the World*. It should not be forgotten that a circle of associations formed the matrix of socio-cultural life in the modern era. And, accordingly, the nature of the issuing body was a circle of like-minded people. Funding came from the well-known patron of the arts, Maria Tenisheva, and the ideological point of view was provided by Nietzschean super-humanity. Published by *World-scholar*s, the magazine fits perfectly into the modernist press system. The *World Science Society* was located on the 1st street of Vasilievskiy Island in St. Petersburg and used an observatory located on the roof of the Tenishev School, one of the best in the country: the school was created by Maria Tenisheva's spouse, prince Vyacheslav Tenishev. The society set itself such a global goal that it would be enough to sustain it for the entire 20th century: with visions such as the popularization of interplanetary flights. Among the members of the society was the "father of Russian cosmonautics" Konstantin Tsiolkovsky, as well as the author of a series of entertaining books on the physical and mathematical fields of knowledge, Jacob Perelman. The society was so popular that its branches arose even in the provincial school environment.

In the editorial introduction of the first issue, the following positions were declared:

> Our magazine will not exist for those amateurs for whom science is entertainment, but for friends of knowledge—those amateurs, imperceptible, but who's numerous works often brought great benefit to science in its forward movement and from which often nominated names that later occupied a prominent place in science. We warmly believe and hope that such friends, by what is inherent in them as real amateurs of the broad view of science in general, will understand and support us because our aspirations are their aspirations and our business should become their business. Concordia parvae res crescent! (Izvestiya russkogo obshchestva lyubitelej mirovedeniya 1912, p. 2)

"The breadth of the view on science in general" was manifested in the magazine in its gravitation toward a "commonwealth of sciences"—the central achievement of the Russian-Soviet popular science press in the 20th century. In the Russian tradition, this magazine can be considered the most specialized within the popular science category, which is due to the specifics of the publisher, that is to say, the scientific circle. In this case, the degree of specialized knowledge is correlated with the material representation, as is evident in such examples of the Western popular science press as the *Scientific American* (established in 1845).[11] For the

[11] In 1980, the *Scientific American's* circulation reached more than 7 million copies a year. And still today the magazine remains at this very high level.

Scientific American in the late 19th and early 20th centuries, a significant thematic diversity of materials related to different countries is evident, indicating the wide range of the magazine's audience. For example, in just one issue of 1900, there were articles on *The New Russian Battleship "Petropavlovsk", Alexandre III. Memorial Bridge, Paris, Death Valley, California, A Pair of Curious Relics at Aberfoyle, Scotland* (*Scientific American* 1900). The *Scientific American's* success for such a long period clearly indicates that it invariably catches the "nerve of time". The process of the journal's formation coincided with the emergence of its mass audience. Therefore, the basic model of the journal combines the popular science component with the elements of a travel agenda. This model should be considered universal.

Abraham Moles correlated science popularization with informal adult education:

> It's about science popularization, which more and more resembles some huge adult education system, on the one hand, overlapping traditional education, and on the other hand, continuing it (up to the coverage of people of the most advanced age, because the old ideas about the world, in which they live, are inadequate). The origin of this system is very interesting since it arose in 1910—1930 from the "scientific and entertaining" literature (*Popular Mechanics, La Science et la Vie, La Nature*), although, of course, there are more ancient predecessors. The fact is that some scientific discoveries can be easily explained to a relatively small number of people who are able to assimilate scientific material and receive from it the same pleasure that others experience from solving a crossword or reading scores. <...> Science popularization is a mosaic culture product <...> at the same time; it is one of the forms of reaction to the superficial nature of knowledge, necessary to define "mosaic thinking". <...>
>
> Thus, science popularization involves two different cycles. One of which resembles a cycle of scientific publication, and another—a cycle of large periodicals: their connection occurs at the publication level. Literary talent combines here with scientific talent, which extremely limits the number of creators in this field. In this case, creativity is manifested not in scientific discovery not in the influence of art. It is the creation of meaningful messages on the topic of some discovery, messages that could be absorbed by any person who has some minimal culture and the ability to make an effort in the broadest sense of the word.
>
> This discovery of the discovery itself is sometimes no less difficult than the discovery of the original. The best proof of this is that the techniques found by one author are repeated many times by others, passing from generation to generation in order to finally take a place in lectures and textbooks (Moles 2008, pp. 238–243).

In this important quotation, let us single out the theoretical and historical component. In accordance with the first, the set of science popularization methods is very limited and universal. In accordance with the second, the system of "scientific and entertaining literature" was formed at the beginning of the 20th century. We fully agree and rely on both theses. However, in Russia, the system of public education was aimed at folk and family (including children's) reading. This feature is caused by the wide educational character of science popularization in Russia.

In response to the demand of Russian readers at the turn of the 19th—20th centuries, mass illustrated publications were predominantly magazines and almanacs with closely related themes, i.e. nature and travel. The closest prototype of nature journals were calendars, where for centuries practical information and folk wisdom had been accumulated. One of the most famous folk calendars—the "Universal Russian Calendar"—was published by the famous book publisher Ivan Sytin (it was first presented at the Nizhny Novgorod exhibition in 1884, and by 1916 its circulation exceeded 21 million copies). Along with desktop calendars, Sytin—with the support of Leo Tolstoy—began to produce tear-off calendars. At the same time, since the middle of the 19th century, the publication of calendars became a popular business, coupled with the publication of educational, scientific, artistic, agricultural and other literature.

The cultural memory of at least several generations has preserved the name of Peter Soykin (1862—1938). Soykin was a unique and versatile publisher engaged in the dissemination of scientific knowledge in an exciting form. From scientific material he made an aesthetically attractive and commercially successful product designed for the widest audience and, more specifically, for family reading. The "children of the Silver Age" generation was raised on Soykin's magazine (as later Soviet children were on the magazine "Young Naturalist"). Actually, since then the family audience seems to remain the most universal for the Russian popular science press. The rapid development of physics, astronomy, biology, and the unprecedented interest in them in all sectors of society, largely determined the publishing path of Soykin. His pre-eminent weekly magazine was entitled: *Priroda i lyudi: illyustrirovannyj zhurnal nauki, iskusstva i literatury (Nature and People: an Illustrated Journal of Science, Art and Literature)* (1889—1918). The content of the magazine was presented in a plot-fictionalized format; and we might consider the magazine as a super-text that relates to a meta-text, providing the reader a key for reading other materials such as adventure literature (Soykin regularly published adventure novels). The mission of the magazine was to enable readers to become acquainted with nature and people of other countries; with the latest discoveries of science; with the secrets of the underworld, the abyss of the ocean and the boundless spaces of the heavens in a fascinating form. The magazine also published the biographies of great scientists, famous travelers, popular essays on all branches of natural science, articles on earthquakes, hurricanes, volcanic eruptions, waterfalls, caves and glaciers, and new trips and discoveries were reported. In full accordance with the actual practice of illustrated weeklies, in addition to *Nature and People*, a number of supplements were published. These supplements were a widely available encyclopedia of various sciences, including A. Brem's *Life of Animals*, M. Faraday's *The Chemical History of the Candle*, as well as a library of adventure novels. In total, Soykin released over 80 million books, 32 magazines, 7 newspapers, and in many cases he acted as the editor. Peter Soykin's case is an example of a self-taught enthusiast, an amateur who managed to combine the service of science with significant public benefit and entrepreneurship.

Thus, summing up the first part, we draw the following conclusions. It seems productive to distinguish three basic principles that determine a specific historical model of science mediatization. The first two have an international character; the third is more applicable to Russia as a logocentric country. So, firstly, this is the level of development and publicity of science itself. Secondly, it is a well-coordinated activity of the society's institutions connected with general education and enlightenment. Thirdly, are the general processes taking place in the media (including the convergence trend of modernity), and the level of social responsibility typical of the press.

It was at the turn of the 19th—20th centuries that the media, in a global context, became the leading channel for science mediatization (not always directly connected with scientific societies). This feature is explained, on the one hand, by the emergence of a mass audience, and on the other hand, by the commercialization of the publishing process itself. From the modern terminological point of view, the characteristic model of the popular science press at the turn of the 19th—20th centuries can be described as trans-media, or hybrid media, with the organic combination of education, enlightenment, and entertainment functions. At the same time, the most important task of this press segment was to develop the audience's understanding of the world. Probably, such a model can be considered optimal. It has produced well-coordinated work for the most important social institutions of modern society: education, enlightenment, and journalism.

From a comparative point of view, as applied to the beginning of the 20th century, the following typological of the popular science press can be distinguished:

1) Narrow special (monothematic) editions, determined by the general process of specialized scientific knowledge. These were devoted mainly to actively developing industries such as technology, medicine, aeronautics, etc.
2) General science magazines, including self-education magazines and universally-specialized magazines. In Russia, the establishment of the universal popular science magazines as an independent and commercialized segment of the press occurred at the turn of the 19th—20th centuries. In the West, this type of magazine appeared earlier, from the beginning of the 19th century. In addition, there is a symptomatic difference in the types of content featured. In Russia, this content included social and humanistic elements, whereas the Western type of popular science magazines should be qualified as universal-specialized. A typical example is *The Magazine of Popular Science, and Journal of the Useful Arts* (1837). It contains only three sections in the table of contents, although inside the journal the material contained in the various sections are freely alternating. The first section—"Original Papers"— includes: "A Popular Course of Geology", "A Popular Course of Chemistry", "A Popular Course of Astronomy". Then follow non-typed subsections of popular science articles (such as "A new French Lamp"), review ("A brief Sketch of early English Scientific Literature". The second and third headings are entitled: "Analysis of Scientific Publications", and "Miscellanea". The last column covers current scientific and near-scientific life: it contains notes on scientific societies' activities and scientific meetings, and even excerpts from "An Act in Addition to the Promote the Progress of Science and Useful Arts" and "The Bill in the United States Legislature, which we recently noted, has become a LAW" (The Magazine of Popular Science, and Journal of the Useful Arts 1837, p. 473). The sheer volume of this journal is considerable; the materials presented are accompanied by tables, formulae, and drawings, with an index at the end. This magazine is focused on the natural sciences; its popular character is provided mainly by the variety of materials rather than the narrowly specialized nature of general science. However, it is calculated to appeal to a scientifically prepared reader. This is why the publication is described

simultaneously as a "magazine" and a "journal". "Useful arts" during the Victorian era meant technique; in Russia, the arts, including the applied arts, belong to the cultural sphere. This difference in terminology is significant as it illustrates more fundamental differences in the approach to scientific popularization.

3) Quality mass publications, presented by magazines and almanacs, with nature (in a broad sense) and travel as thematic dominants. The huge popularity of these kinds of publication among readers was determined, on the one hand, by the effectiveness of the education and enlightenment system; and, on the other, the development of journalism.

References

Balashova, Yu.B. 2007. *Shkol'naia zhurnalistika Serebrianogo veka* [School Journalism of the Silver Age]. St. Petersburg: Saint Petersburg University Publishing House.

Beliy, Andrey. 1989. *Na rubezhe dvuh stoletiy. Vospominaniya* [At the turn of two centuries. Memories]. Vol. 1. Moscow: Hudozhestvennaya literature.

Berdyaev, Nikolay. 2005. *Filosofiya svobodyi* [Philosophy of freedom]. Moscow: AST.

Graeme, Donald. 2013. *The Accidental Scientist. The Role of Chance and Luck in Scientific Discovery*. London: Michael O'Mara Books Limited.

Ivanov, A.Ye. 1999. *Studenchestvo Rossii kontsa XIX—nachala XX veka: sotsialno-istoricheskaya sudba* [Russian students in the late 19th and early 20th centuries: social and historical fate]. Moscow: Russian Political Encyclopedia.

Izvestiya russkogo obshchestva lyubitelej mirovedeniya [News of the Russian Society of Amateurs of the World]. 1912, 1.

Kokurina, Elena. 2018. Mir soznaniya i soznanie mira [The world of consciousness and the consciousness of the world]. In: *V mire nauki* [In the World of Science], 5–6. Accessed July 2018. https://scientificrussia.ru/articles/mir-soznaniya-i-soznanie-mira-v-mire-nauki-5-6-2018-g

Moles, Abraham. 2008. *Socio-dynamics of culture*. Moscow: Publishing house LCI.

Muravyeva, Irina. 2004. *Vek moderna: panorama stolichnoy zhizni* [The modern age: panorama of life in the capital city]. Saint Petersburg: Pushkin Foundation Publishing House, 2.

Foucault, Michel. 2004. *The Archaeology of Knowledge*. London: Routledge.
Scientific American. 1900, May 19, Volume 82, Issue 20.
The Magazine of Popular Science, and Journal of the Useful Arts. 1837. Volume the third. London: John W. Parker, West Strand.

PART II

THE SOVIET EXPERIENCE OF SCIENCE MEDIATIZATION

CHAPTER ONE

SCIENCE COMMUNICATION AND ADVANCED SCIENCE IN THE EARLY SOVIET PERIOD

After Bolshevik revolution of 1917, in many spheres of public life, the cultural succession continued to function paradoxically. In the field of culture, modernist trends were clearly preserved until about the mid-1920s, which was determined by the powerful potential of modernism. At the same time, scientific continuity explicitly manifested itself in the Soviet Union over a longer period of time than was true of cultural or social institutions. The cardinal change of the scientific paradigm occurred in the Soviet period only after the Second World War; and science in the USSR in the 1920s—1930s developed under the direct influence of this breakthrough in advanced science at the turn of the 19th and 20th centuries. This kind of specific continuity, which took place despite the massive departure of middle-level scientific personnel (privat-docents, largely settled in the West) is explained, in our opinion, by the following factor. Among the representatives of the "first wave" of scientific emigration (in total in the 20th century there is usually considered to have been three waves of Russian emigration), such great scientists as aircraft designer Igor Sikorsky, and TV founder Vladimir Zvorykin left Russia.[12] However, the seemingly paradoxical fact of the preservation and reproduction of scientific paradigms after the shock of 1917 can be explained by three further factors.

First, from the beginning in the young Soviet state there was no own scientific policy, and, as in the case of the construction of other state institutions (for example, censorship), the model used in the Russian Empire was taken as a basis. Official science under the aegis of the Academy of Sciences quickly began to cooperate with the Bolsheviks (as did the Russian Orthodox Church, despite the church split). Thus, in 1919, the leading Russian mathematician Vladimir Steklov (educated in an

[12] According to contemporary statistics, from an economic standpoint, every single Zvorykin leaving the country can be estimated as a loss for the country of 20 GDP (gross domestic product).

environment with democratic traditions) became the vice-president of the Academy of Sciences, and in many respects, thanks to his efforts, the Imperial Academy of Sciences was reorganized into the USSR Academy of Sciences. Henceforth, world-renowned scientists began to associate the revolution with the general process of democratization and the establishment of a relationship between science and society. In this regard, we should note the symbolic story of the famous celebrity scientist Kliment Timiryazev. In Andrey Beliy's memorial address, he characterized him as a truly modern man: "who was able to imbue the power of feeling in every step <...>" (Beliy 1989, p. 432). Nevertheless, Timiryazev not only accepted the new regime but also actively supported it. In 1920, a collection of articles were published under Timiryazev's editorship under the landmark title: "Science and Democracy". In a brief preamble, Timiryazev directly identifies "scientific truth" and "socialist truth": "... these pages are connected by a common desire for scientific truth and to ethical, socio-ethical, socialist truth" (Timiryazev 1920, p. 5). The collection contains Timiryazev's arguments about popularization as such: "The role of science popularization, we can say, was fully recognized only by the great scientists of the middle of the last century (Faraday, <...> Darwin <...>, etc.). And the importance of popularization is growing with the growth of democracy" (Timiryazev. 1920, pp. 11–12). The contribution made by the popular works of Timiryazev himself is usually assessed as follows:

> In Russia, Kliment Arkadyevich Timiryazev was the first great scientist who proved that it is possible to present the results of his research in such a way that they were understood not only by specialists but also by all literate people. He was the first to make a path down which many remarkable scientists then went. Without the "Life of Plants" and other well-known books by Timiryazev, probably, books of such scientists as A.E. Fersman, V.A. Obruchev, and others, books that have become classics of science popularization, would not have appeared (Razgon 1983, p. 126).

The revolution offered upward social mobility for talented representatives of the lower social classes; people who were self-taught. Before the revolution, the aforementioned "father of Russian cosmonautics", Konstantin Tsiolkovsky, had some difficulties with the recognition and publication of his scientific works, which seemed utopian; however, in the USSR, although not immediately, his works began to be considered fundamental. This is a reminder that scientific breakthroughs often arise in those areas that are generally considered fantastic, marginal, pseudoscientific or, indeed, utopian.

The cultural life of a social matrix is formed by a variety of institutional and non-institutional associations (on Russian soil, primarily a "circle" rather than of a "club" type). After the revolution, on the one hand, pre-revolutionary scientific societies continued to operate (for example, the Russian Geographical Society, the Society of Lovers of Russian Literature); and on the other, new avant-garde communities were created. In the designated sphere, the influence of modern culture was also manifested, with its cult of professional and creative cooperation (based on the idea of "conciliarism" and "life-building"). So, in the second half of the 1920s, student Dmitry Likhachev (later to become a well-known philologist and public figure) took an active part in the joke science society: "The Space Academy of Sciences". Participants in this self-educational student circle in carnivalesque forms engaged in "fun science": they made reports, received the appropriate masquerade "chairs" ("Melancholic Philology") and even sent a congratulatory telegram addressed to the "Academy" on behalf of the Pope. It is noteworthy that the carnival element was an echo of the "Russian cultural Renaissance" of the early 20th century, which formed due to the hybridization of new social realities in the highly original cultural environment of the 1920s, and prevailed over the compensatory and educational function of such communities. Centuries earlier, students of Moscow University, the future "archival youths", created a circle where "we had evening talks that lasted far beyond midnight, and they proved to be much more fruitful for us than all the lessons that we took from the professors. <...> We owe this conversation to very many people both scientifically and morally" (Koshelev 2001, pp. 131–132). As for the burlesque "Space Academy", the consequence of this carnival was the arrest of "academicians" in February 1928. The first scientific work of political prisoner Likhachev, on the games of criminals, was published in 1930 in the camp of political prisoners' magazine "Solovetsky Islands". It is noteworthy that a literary and theatrical circle operated at the "Solovki Special Purpose Camp" at the turn of the 1920s and 1930s; a series of camp magazines were produced, and there was an acting scientific society. From the mid-1920s a single political trend was clearly manifested; due to the unification of various aspects of social life, there was a cultural and scientific process of division into two main streams: official and unofficial, the latter only becoming apparent during the Khrushchev's "thaw" period. Entire branches of knowledge, such as genetics, or formalism were prohibited. In Grigory Pomerant's philosophical book, "Notes of the Ugly Duckling", he brilliantly describes how an intellectual environment was encountered in "correctional labor camps" (the Gulag system). He gives an account of the peculiar phenomenon of the camp "sharazhka" (meaning: "prison-style construction office"), where the designer of rocket and space technology, Sergey Korolev, designed the first manned spaceship. It was in this same spaceship that Yury Gagarin made his space flight on April 12, 1961.

Throughout the Soviet period, a significant role in the matter of scientific education belonged to specific socio-cultural institutions, which, in turn, were exponents of the characteristic Soviet orientation for organizing science and enlightening a wide audience. In the early 1920s, three prominent centers operated in Petrograd: the House of Arts, the House of Writers and the House of Scientists; combining the physical survival of the intellectual elite, the task of party control over the intelligentsia, and the organization of their everyday life (food and shelter) with self-organization and a wide range of cultural and educational activities. Still functioning today, the *House of Scientists* in Soviet times earned a reputation as a subcultural and ideological niche.

Thanks to the activity of leading scientific and cultural forces not only continuity, but also a connection with Western science was achieved. Conversely, and at the same time, within the middle section of scientific workers a corresponding gap began to appear, thus:

> On May 25, 1931, the Politburo of the Central Committee of the All-Union Communist Bolsheviks Party considered the question "Of the Production and Technical Propaganda Formulation". <...> It was decided that "for the successful development of industrial and technical propaganda, both oral (lecture) and literary, <...> a department of production and technical propaganda has to be organized <...>". Already on August 13, 1931, N.I. Bukharin, addressing the workers, stresses: "One of the most acute tasks that have placed upon us with all the force of gravity is the lack of engineering and technical personnel, the shortage of scientific workers. <...> We need to attract the best of our (and foreign) forces, to develop the means of producing first-class technical literature" (quot.: Vaganov 2014, pp. 96–97).

Indeed, if we turn to Andrey Platonov's prose legacy, the heroes of his stories of the 1920s were also Western technical experts.

Secondly, the international pathos of the Russian Revolution at first helped to develop an interest in Western culture and science. Symptomatic was the large-scale series "World Literature", an initiative of Maxim Gorky. The ambition of "World Literature" was to solve the important task of publishing the world classics heritage with detailed comments for a new, mass audience. Similar in importance were the scientific projects themselves—scientific series: "Academia" and "Scientific Heritage". The interest in science was also shown in the early 1920s by "Russian Berlin", where publishing activity resembled the corresponding practice of the Silver Age. "Epokha" Publishing House published the magazine "Conversation: The Journal of Literature and Science" (1923—1925), with a diverse scientific department,

> …introduced at the insistence of M. Gorky <...> It was typical of Gorky's "faith in science" to involve in the "Conversation" materials such as "The Primitive Population of Europe" by Prof. Brown, "X-ray" by Prof. O. Wiener (translated from the manuscript), or "Fundamentals of the Modern Doctrine of Heredity" by Dr. G. Werner, and even "Fundamentals of Radiotelephony" by Prof. Harry Schmidt. The philosophical sketch of L. Ulvig's "Miracle in Science", which was published in the "Conversation", answered in this connection some deep principles of the Gorky worldview. A great deal of space in the magazine was devoted to the history of literature (essays on Byron, Goethe, and French, English, German and American literature) (quot.: Tolstoy 1993, pp. 135–136).

In the non-political "Conversation" (which, nevertheless, was never allowed to enter Soviet Russia), science was included in the typical encyclopedic journal literary form.

And, thirdly, in accordance with the ideological guidelines of the Soviet state, the educational level of the population had to be brought closer to the scientific one, since, as we know, Marxism was interpreted as a purely scientific worldview. With a poorly educated population, it is impossible to build communism. It should be noted that significant results were achieved along this path (one of the first and most important ones was a successful campaign to eliminate illiteracy). With direct state participation, the organization *Proletkul't (Proletarian Culture)* was established and existed from 1917 to 1932. The purpose of this organization was to introduce culture to the masses. *Proletkul't* was a cultural, enlightening, literary-artistic, and truly mass organization. For ordinary workers, "Komsomol" members and young communists; libraries, literacy courses, mobile theaters, and creative studios were organized. Filmmaker, Sergey Eisenstein, became a graduate of the theater studio. Recognized as one of the greatest directors in the history of world cinema, Eisenstein began his career in the troupe of the First *Proletkul't* Theater. The designation for the people's enculturation was accompanied by the development of active and original modes of science communication. Let us note that the founder of the Soviet state Vladimir Lenin believed in the objectivity and the scientific nature of journalism as such, one capable of forming a communist worldview. Therefore, during the Soviet era, the media was described as "mass agitation and propaganda", where the most important role was assigned to scientific problems. In Soviet times, almost every editorial board had a department of science and technology. Throughout the Soviet era, popular science journalism was closely associated with Soviet science and, at the same time, manifested a certain freedom from ideological dictatorship. From such a totalitarian ideology it was protected by the entrenched and largely reasonable idea that pre-revolutionary popular science magazines had served as a legal channel for the spread of Marxism.

This is why some publishers-educators of the previous era, such as Peter Soykin, were given the opportunity to engage in more or less unhindered activity during the 1920s. So, the "cooperative publishing house of P. P. Soykin and I. F. Afanasyev," during the 1920's, published *Mir priklyucheniy. Illyustrirovannyiy zhurnal povestey i rasskazov (World Adventures. Illustrated Magazine of Tales and Short Stories)* (Leningrad, 1922—1930). This magazine continued the diverse and high-circulation series of pre-revolutionary publications by Soykin, more precisely it inherited the same monthly illustrated magazine, "The World of Adventures", which was published during 1910—1918 as an appendix to Soykin's central innovation—the "Nature and People" magazine. The post-revolutionary *World of Adventures* was a typically prosaic almanac, with translated literature (novelties and classics), humor, and game feedback (tasks, riddles, charades, rebuses, etc.). This miscellany mostly developed the classical traditions of the Russian enlightenment, which was inseparable from scientific popularization. Such a brand of Soviet journalism as the *Novyiy mir (New World)* magazine (which played a landmark role in the liberalization of public relations during Khrushchev's and Gorbachev's times) arose in 1925 as a "literary-political and scientific magazine".

The experimental external context (political-ideological, socio-cultural, and paradigmatic) contributed to the revolutionary upsurge of Russian science, primarily in its humanistic aspect. The 1920s became a time of advanced humanist science, developed according to the avant-garde scenario. Its nomination was a direct consequence of the formation of scientific paradigms in the early 20th century. Advanced humanist science entered the public sphere mainly through the literary press and, to a lesser extent, through specialized scientific publications. Thus, the perspective of the formalist method in the literary criticism of Yury Tynyanov, concerning the abrupt, discontinuous character of literary evolution, can be directly correlated with concepts of biological science, more precisely: the value of mutations in biological evolution. In fact, in Tynyanov's work, "genre as a gene" is specifically discussed (Tynyanov 1997, p. 511). More recently, another representative of the formalist school, Victor Shklovsky, writes the following:

> Like his contemporaries, Victor Shklovsky was fascinated by modern science, from Einstein's theory of relativity to the quantum and wave theories of light and Nikolai Lobachevsky's conception of a non-Euclidian geometry that doesn't accept the central axiom that parallel lines cannot meet (Boym 2017, p. 4).

The scientific-utopian coloring of the epoch determined such aspirations as the quest for immortality for the political elite (expressed in Michael Bulgakov's prose), the "philosophy of Russian cosmism", combining a naturalist and religious-philosophical orientation (which was influenced by the scientific utopianism of A. Platonov's literary world). The "Philosophy of Russian cosmism" had religious and philosophical roots, and arose on the basis of a combination of naturalist and humanistic ideas, specifically reflected on Russian soil in the idea of a cosmic "all-unity". The great naturalist and "nationalist" Vladimir Vernadsky (1863—1945), associated with the "cosmists", constantly returned to the thesis that: "the consequences of the newly-discovered areas of scientific facts caused a simultaneous change in the very foundations of our scientific knowledge of the environment <…>" (Vernadsky 2014, p. 69). Vernadsky developed the concept of the inevitable transition of the biosphere to the "noosphere"—the sphere of the rational and creative comprehension of the world that unites mankind. In fact, Vernadsky predicted the advent of the information era, and he considered scientific knowledge to be its main developmental factor. Such modern scientific and para-scientific trends—such as telecasting and cloning—were also included in the sphere of attention of Russian cosmists. Self-taught inventor, Konstantin Tsiolkovsky, developed projected spacecraft as a practical application of his "cosmic philosophy".

Science claims to create a matrix of the 20th century human consciousness, which is why the 20th century as a whole has devoted so much attention to science fiction. The very nature of the science fiction genre is closely connected with social utopianism. It is a well-known fact that, after Orson Welles' radio show on Herbert Wells' novel *The War of the Worlds* in the late 1930s (the novel was published at the end of the 19th century), US citizens succumbed to mass panic, awaiting the capture of the Earth by the Martians, and Sci-fi stories, which seemed unrealizable at the time of their publication, now began to be perceived as prophetic. Thus, the detective plot of the famous Soviet science fiction writer Alexander Belyaev's novel, *The Head of Professor Dowell* (1925), connected with the experience of recreating the functions of the human brain separately from the body, finds analogies in modern experiments on head transplantation. After the publication of Belyaev's novel, a public discussion was organized at the First Leningrad Medical Institute. Contrary to the widespread view of the impregnable "iron curtain", contacts between the USSR and the Western countries existed in 1920—1930. The same Herbert Wells visited the USSR several times, where he met not only with the Soviet state leaders but also with Alexander Belyaev, whom he regarded very highly. It is noteworthy that Belyaev's science fiction novel *Struggle in the Air* (1927) was diligently studied during the Cold War by the Central Intelligence Agency since a hypothetical war between the USSR and the USA was described there.

During the Soviet era, the official cult of scientific and technological progress was expected to replace the repressed religious consciousness. The authorities not only supported certain scientific developments but were inclined on occasions to treat important scientists with a certain mystical reverence (as, for example, the fate of Peter Kapitsa testified: during the time of his disgrace he nevertheless had the opportunity to continue physical experiments, albeit secretly in a garage). Concerning the limitations of speculative utopias, not sufficiently rooted in the soil and traditionalist folk spirit, Andrey Platonov wrote, in a journalistic note under the title *About Love*:

> ...we will give the people science in exchange for religion. This gift will not console people. Science in the modern sense has existed for only 100—150 years, religion—tens of thousands of years. What is stronger and what is more deeply embedded in the inner man? Scientists themselves, the very leaders of science, were almost all believers. Science is still at such a low level that it cannot be a leader of a person, a force that stands above it. Life is still wiser and deeper than any thought, the elements are considerably stronger than consciousness, and all attempts to replace religion with science will not lead to the complete victory of science. People need a different, higher, more universal concept than both religion and science (Platonov 2011, p. 11).

In the situation where the traditional socio-political discourse in Russian society fell into decline, it was steadily replaced by literary discourse. Scientific problems can also create a public sphere, forming a kind of meta-text. If in the 19th century the object of artistic reflection was Russian history and the problem of power and social interaction, then in the 20th century fiction concentrated more on issues of scientific and technological progress. Thus, the famous poet, Osip Mandelstam, wrote such articles as: *Around the Naturalists*, and *To the Problem of Darwin's*

Literary Style. In this connection, the creative evolution of satirist Michael Zoshchenko is highly indicative. In the early 1930s, he was striving to form a less satirical more positive image, based on the modern scientific context. One of the central ideas in Zoshchenko's mixed-genre books, such as *Vozvraschonnaya molodost' (Returned Youth)*, and *Pered voshodom solntsa (Before Sunrise)*, consists in actualizing the acute need for self-knowledge in a person. A crucial role in this process is assigned to objective scientific knowledge and on the research of the writer's contemporaries, e.g. Ivan Pavlov on higher nervous activity or Vladimir Bekhterev on the activity of the brain. In full accordance with the current scientific context, Zoshchenko publically formulates and artfully proves that the brain regulates and controls the energy and the psyche of man.

Several years ago, interesting transcripts of the public discussion of Zoshchenko's story *Returned Youth* were published. They demonstrate the undoubted interest of the scientific community in this writer's work. The academic, Ivan Pavlov, began inviting Zoshchenko to the scientific society meetings at the Institute of Physiology.

> From March to May 1934, a broad discussion of this book developed, both in literary circles and scientific ones. It is curious that these numerous discussions were caused not only by the interest in the famous and popular writer's work, who overcame a certain creative boundary; they were also inspired by another ideological campaign aimed at "merging" science and literature. In one of the reports on the activities of Leningrad writers, in which scientists from the Academy of Sciences took part, it was reported that "about a year and a half ago, Academician N.Ya. Marr expressed the extremely important idea that in some remote perspective the ideology of science can merge with the ideology of fiction". Therefore, the story of M. Zoshchenko was recommended to be considered along with such books as "The People of the Stalingrad Tavern Factory", <...> the collective work of writers about the White Sea-Baltic Canal.[13] However, this context turned out to be very far from the story of M. Zoshchenko, and all the discussions took place only within the framework of medical discourse (Ezhegodnik Rukopisnogo otdela Pushkinskogo Doma 2013, p. 829).

At one of the meetings a representative of the scientific community made the following important points:

[13] The White Sea—Baltic Canal (now rarely used) was built by the hands of political prisoners of the Gulag; Soviet writers were obliged to sing this construction of communism.

> Speaking about Zoshchenko's book, it seems to me that it would be necessary to raise in general the question about that role, or about the relation that literature, in general, can manifest to science. The fact is that for a long time everybody knows that good literary works served as material for science, that Dostoevsky is considered an excellent psychologist to the extent that even science did not reach. We can say that physics can explain the structure of the atomic nucleus much better, physics answers questions that the philistine does not imagine, but physics cannot explain the question of why there are sugar and protein. Therefore, between life and science, there is a big chasm, and literature is filling this gap. In general, in those days when science was still in its infancy, who then answered all the questions posed by man? The literature answered these questions (Ezhegodnik Rukopisnogo otdela Pushkinskogo Doma 2013, p. 849).

It is also not accidental that it was the Soviet era that championed the task of pursuing a "literature of fact", in contrast to classical fiction. Thus, during Soviet times, fiction was the most important channel for science mediatization. In the West, similar attempts were also made. A man of the 20th century is expected to possess a scientific worldview, and not in contrast to or opposed to the philosophical and poetic perception of nature. Such maxims are quite appropriate in types of publication such as the almanac:

> *An Almanac for Moderns* was designed for reading day by day if the reader so pleases. But the pages were not written, actually or in outward pretense, as an ordinary diary, even a Nature record of some particular year. The *Almanac* was conceived, rather, as the common experience of any year, with its impact of seasonal sight and sound, and its glimpses of the birds' ways and the slow self-robing and disrobing of the woods. Still more, it is intended to convert the overtones of these events as they ring upon a modern ear with a scientific training. By overtones, I am far from meaning (to continue the musical figure) a transposition of the themes of Nature into an ethical key. An overtone is something that you cannot force; you hear what you hear, and in the slighter pages of the book the reader will understand that the author heard nothing but the melodic passage itself and did not try to raise an echo and call it a phrase dropped from the spheres
>
> The biological viewpoint in this book has been deplored as materialistic. If the natural sciences were verily thus, would we not expect to find their disciples cynical and grasping, devoted to their own appetites and glory? The lives of the naturalists, their reputations in the world today, are quite otherwise. I may praise their men since I do not consider myself one of their number; living scientists, or dead ones who have left the world their legacy, have given me of their best. I have been privileged to enjoy the training of a naturalist without ever having practiced more than

modestly. All that is true in these pages I owe to my reaches. Only its shortcomings are mine (An Almanac for Moderns 1938, pp. 5–7).

It is noteworthy that during the periods of liberalization in the USSR, environmental problems began to come to the fore in the public agenda.

Popular science journalism between 1920—1930s occupied a secondary position in Soviet Russia in relation to literary scientific education and various oral channels popularizing knowledge (associations, public discussions, etc.). The "golden age" of popular science journalism in the USSR is associated with the period of 1960—1970s when almost every Soviet family subscribed to a popular science magazine and watched popular science and educational programs on television.

Let's repeat one of the key themes: the tendency to unite naturalist and humanist themes mark the Russian-Soviet tradition of the popular science press, highlighted against the background of analogous Western practices. Consider one of the most popular journals in the pre-revolutionary period, *Vestnik znaniya (The Herald of Knowledge)*, by V. Bitner. *The Herald of Knowledge* was resumed in 1922 and was published by the publishing house "P. P. Soykin"; his editors were first V. M. Bekhterev, and then the historian S. F. Platonov, the Soviet historical periodicals creator. The editorial work of these two outstanding scientists reflects the general approach characteristic of the Soviet era when the propagation of knowledge was imputed to the scientists themselves. As an example, on the basis of a selective principle, we analyze one of the issues of *The Herald of Knowledge* from the mid-1920s. Attention is drawn to the genre-thematic diversity of the magazine, something inherited from the previous era. The most relevant scientific and social problems were reflected in such articles as: "New experiments of rejuvenation in the USSR", and "Sexuality and crime". The humanistic component was represented mainly by the literary critic A. A. Rashkovskaya's article "The Newest Quest in Modern Literature" (she was well known for her opposition to the formalists). From that time onwards it was typical for popular science magazines to appear. One of them is devoted to local history (in this case—"Along the Native Land"), another one is press reviews (here—"New in the Press"). The international scientific context, presented in a sensational way, was concentrated in the final heading of the magazine: "From All World Corners". It included such notes as: "World Records of 1925", "Alcohol From Acorns", "Curios of the Calendar", and "The Most Cultured Country in the World" (by the way, this was considered to be the Canadian province of Ontario, where 3 million people had 460 public libraries) (Vestnik znaniya 1925).

For comparison, in the French journal of the same time and from the same typological group, under the iconic name: *La Science et la Vie (Science and Life)*, the humanistic component was totally absent. At the same time, the journal is sufficiently specialized (technical), and its orientation toward a mass-specialized audience is achieved by following the path of rich illustrations and abundant advertising (*La Science et la Vie*. 1929). A close, though not identical example, typifies American popular science press. For example, *Popular Science Monthly* (founded in 1872) published a great deal of illustrated scientific information and was committed to this very practical (pragmatic) task. A significant amount of material was focused on what one can do oneself at home—following the recommendations outlined in the magazine. See, for example, its content: "Contents For January 1939: Features and Departments (Automobiles, Aviation, Health & Hygiene, Helps For Home Owners)" (*Popular Science Monthly* 1939).

So, the intention of the magazine is enlightening-commercial, promoting presented mechanisms and tools. Similar pragmatism was also characteristic of the Soviet popular science magazines during the second half of the 20th century. However, in the USSR such pragmatism was not at all caused by the forces of the market and marketing. The fact is that in a planned economy there was such a phenomenon as a deficit. The information from the heading "do it by yourself", really useful in everyday life, was mainly borrowed from popular science magazines.

In the Soviet popular science magazines of the period under review, a ritual ideological component was mandatory, as was true of the whole Soviet press (all the editorials were filled with information about the next plenum of the Communist party congress, panegyrics to the Soviet leaders, etc.). Despite this, these magazines stand out against the background of their very different European and American analogs. They were distinguished by their prevailing attitude toward the development of the worldview of the audience and, to a lesser extent, practical and useful everyday life skills. This was achieved by the convergence of naturalist and humanistic knowledge.

Thus, we will sum up the results of this chapter based on the perspective of science historian Thomas S. Kuhn and his notion of a scientific paradigm shift. Thomas Kuhn identified several signs of such a shift. First of all, in scientific revolutions, the analogy with biological evolution is broken. Then, the conceptual discrepancy between different disciplines increases and specialization grows (Kuhn 1962). From the indicated principles, the scientific revolution in Russia—as well as in developed societies as a whole—happened at the turn of the 19th and 20th centuries. That is, social revolution does not automatically lead to irreversible upheavals in other areas. On the contrary, in the long run, the successes of many branches of Soviet science (not only cosmonautics, but also at a certain stage one of the most authoritative in the world, Soviet philological schools) were achieved both by state support and by the specific action of the mechanisms of scientific continuity. The revolutionary reorganization of Russian society, disastrous for the country as a whole, was, nevertheless, accompanied by the original development of science communication.

References

An Almanac for Moderns. 1938. By Donald Culross Peattie. Including a New Introduction by the Author and Wood Engravings by Asa Cheffets. Printed in Washington for the Members of The Limited Editions Club.

Beliy, Andrey. 1989. *Na rubezhe dvuh stoletiy. Vospominaniya* [At the turn of two centuries. Memories]. Vol. 1. Moscow: Hudozhestvennaya literature.

Boym, Svetlana. 2017. *The Off-Modern.* New York, London, Oxford, New Delhi, Sydney: Bloomsbury.

Ezhegodnik Rukopisnogo otdela Pushkinskogo Doma na 2012 god. 2013 [Yearbook of the Handwriting Department of the Pushkin House for 2012]. St. Petersburg: Dmitriy Bulanin.

Koshelev, Alexander. Kruzhok arhivnyih yunoshey [A circle of the archival young men]. In: *Literaturnyie salonyi i kruzhki: Pervaya polovina XIX veka* [Literary salons and circles: The first half of the XIX century]. Ed. by N.L. Brodsky. Moscow: Agraf.

Kuhn, Thomas. 1962. *The Structure of Scientific Revolutions.* Chicago: The University of Chicago Press.

La Science et la Vie. 1929, № 144. Paris: Imp. Hémery.

Platonov, A.P. 2011. *Publtsistika* [Publicism]. Moscow: Kniga po Trebovaniyu.

Popular Science Monthly. 1939. Vol. 134, № 1.

Razgon, L. 1983. *Zrimoe znanie: O knigah K.A. Timiryazeva i A.E. Fersmana* [Visual knowledge: About the books of K.A. Timiryazev and A.E. Fersman]. Moscow: Kniga.

Timiryazev, Kliment. 1920. *Nauka i demokratiya: sbornik statej, 1904—1919* [Science and Democracy: a Collection of Articles, 1904—1919]. Moscow: Gosudarstvennoe izdatel’stvo.

Tolstoy, Ivan. 1993. *Kursiv epohi: Literaturnyie zametki* [Cursive of the Epoch: Literary Notes]. St. Petersburg: Pushkin Foundation Publishing.

Tynyanov, Yu.N. 1997. Literaturnaya myisl'. Almanah II [Literary thought. Almanac II]. In: Tynyanov, Yu.N. *Poetika. Istoriya literaturyi. Kino* [Poetics. History of literature. Cinema]. Moscow: Nauka.

Vaganov, Andrey. 2014. *Spiral' zhanra: Ot "narodnoj nauk"' do razvlekatel'nogo biznesa: Istoriya i perspektivy populyarizacii nauki v Rossii* [Spiral of the genre: From "folk science" to entertaining business: History and prospects of popularization of science in Russia]. Moscow: Lenand.

Vernadsky, V.I. 2014. *Filosofskie myisli naturalista* [Philosophical Thoughts of the Naturalist]. Moscow: Akademicheskiy proekt.

Vestnik znaniya [The Herald of Knowledge]. 1925, № 21—22. Leningrad: "P.P. Soykin" Publishing.

CHAPTER TWO

THE DIVERSITY OF SCIENCE MEDIATIZATION CHANNELS IN THE USSR

After the end of World War II, the organization of science and education on a global scale underwent significant changes. Several factors had become decisive. The scientific and technological revolution made society rethink its ideas about the surrounding world, as well as the role of mankind within it. Another factor was the emergence of the Cold War, dividing the world into two opposing systems. The invention of the nuclear bomb *(the Manhattan Project in the USA, the Nuclear Project in the USSR*) eventually became a deterrent, preventing a possible catastrophic war between the socialist and capitalist camps.

The development of science in the USSR was determined by the enormous attention paid to it by the state. After World War II scientific management began to concentrate more and more in the hands of party functionaries rather than scientists themselves, reminding us of the well-known fact that science in totalitarian regimes is closely related to state ideology.

> However, the relationship between science and Nazism could also be strong and mutually constitutive. Human genetics, anthropology, and medicine, for example, not only provide plenty of cases where Nazism furthered the developments of the sciences in certain directions but also illustrate that the sciences created languages, illustrations, evidence and theories that helped create Nazi ideologies and guide Nazi actions (Agar 2012, p. 213).

The problems of scientific and technological progress in the post-war period were resolved in the USSR with a strict account of the international situation and the aggravation of relations between the two military-political blocs within the framework of the Cold War. Having entered into competition with the United States for strategic superiority, the USSR was forced to spend huge amounts of money on the implementation of the atomic project and, a little later on, space exploration. A huge amount of money was spent on the arms race; in 1952 the direct military expenditure of the USSR amounted to almost a quarter of the country's total annual budget. In this regard, the military-industrial complex was placed in an exclusive position. In the first post-war years, the USSR managed to liquidate the US monopoly on nuclear weapons in a very short time. In the summer of 1949, an atomic bomb was tested at the Semipalatinsk test site. In 1953 a new powerful thermonuclear bomb was tested in Novaya Zemlya, one of the founders of which was the academic and human rights activist *Andrey Sakharov*. An academic, *Igor Kurchatov*, was the scientific leader of the whole atomic project. During the same years, Soviet scientists worked on the problem of using atomic energy for peaceful purposes. Even in the years of the "Fourth Five-Year Plan" (1946—1950), a project was developed for the world's first nuclear power plant, which was commissioned in 1954 in the city of Obninsk, in the Kaluga region.

The development of nuclear projects, both peaceful and military, stimulated, in turn, the development of a number of scientific areas, as well as new industries. Nuclear installations began to be used as power plants on ships (the nuclear icebreaker "Lenin") and on submarines. Priority areas of scientific and technical policy were the problems of nuclear physics, missile and aircraft technology, and the chemical industry. In the 1950—1960s, radio engineering and electronics were rapidly developing. It was then that television entered the homes of many Soviet people. During the same period, Soviet scientists *Alexander Prokhorov* and *Nikolay Basov*, and in parallel American physicist Charles Hard Townes, created the world's first laser. Significant results have been achieved in the field of welding and the creation of electric welding equipment. Let us repeat, for the period under consideration the Soviet state paid close attention to the development of science. New academic and branch scientific centers were created. The Academy of Sciences of the USSR assumed the function of coordinating the work of scientific institutions. As such, great scientific potential was concentrated in academic institutions. Among them, it is worth emphasizing the physical institutes named by P. N. Lebedev and S. I. Vavilov: The Institute of Physical Problems, The Institute of Physical Chemistry, etc.

Science and technical achievements were such that, beginning from the mid-1950s, the USSR, together with other advanced countries, entered the era of a scientific and technological revolution. Here is an informal, although accurate description of the value of science in the Soviet society of 1960, given by representatives of the third wave of Russian emigration: writers Peter Vail and Alexander Genis (engaged also in alternative literary criticism and the history of culture):

> Since the country took the course to build communism, the question became increasingly acute: who should build it? To answer this question, the 60th had to find their heroes. <…>
>
> The new great state truth must be based on a solid foundation, not subject to political upheavals. The 20th century reasonably offered science as a foundation.
>
> In the eyes of society, scientists had decisive dignity and honesty. The last one is also sincerity, decency, truthfulness. The era made all these words synonymous and invested in them a worldview.
>
> Twice two must be equal to four, regardless of the principles of the one who believes. After an arbitrary Soviet past, the country was in dire need of an irreplaceable present. The multiplication table possessed the qualities of absolute truth. Exact knowledge seemed equivalent to moral truth. Between honesty and mathematics was put an equal sign.
>
> After it turned out that the words were lying, more confidence was caused by the formulas.
>
> Scientists lived nearby, scientists were ordinary Soviet people. And yet—others. Not for nothing in the newspaper jargon era were they called priests of science.
>
> Society, gradually freed from the belief in the infallibility of the party and government, feverishly sought a new cult. Science was approached in all respects. It combined the objectivity of truth with the incomprehensibility of its expression. Only those who are initiated into the sacraments can serve science in its temples (Vail and Genis 2018, pp. 117–118).

Meanwhile, in Khrushchev's time, a winged phrase appeared: "The peace for physicists, the war for lyrics". Humanistic knowledge had increasingly turned into a servant of politics and ideology. A striking example was Nikolay Marr's "new doctrine of language" (his surname was a shortened from the Scottish *Murray*). Academician Marr started to develop this theory from the beginning of 1920, bringing it closer and closer to Marxism. In Marr's interpretation, language, like society, has a pronounced class nature. It represents a Marxist "superstructure" over a "base"—social-economic relations. Accordingly, under communism as the highest stage of societal development, the most perfect single language should arise. Such a pseudoscientific theory was rooted not only in Marxism but also in avant-garde philosophical and linguistic research, in particular, the ideas of the futurist poet Velimir Khlebnikov regarding a "global language". In this respect, the correlation between language and politics is very characteristic (expressed, among other things, in the Esperanto movement). A single language should provide a single world center and a singular world management system. But, despite its official status, Marr's theory was ultimately discredited by Stalin in his work

"Marxism and Questions of Linguistic" (1950). In fact, we have before us a kind of linguistic "theory of elites". Various "theories of elites" were widely disseminated in the 20th century; one of the interpretations, with indubitable Nietzschean allusions, was presented in the famous novel by Ayn Rand "Atlas Shrugged".

The revolutionary historian, Pokrovskiy, argued that "history was a politics turned into the past". The development of the humanities was conducted under vigilant party control. The discussions on philosophy, linguistics, and political economy that took place in the late 1940s and early 1950s were externally called upon to raise the role of social science, but in reality they pursued the goal of establishing rigid ideological control over it, and launching a struggle against "pathetic cosmopolitanism", "bourgeois adulation" and the "admiration of bourgeois authorities". Unlike the humanities, social science did not develop in the USSR at all; and nor did genetics. Genetics confirms the thesis of heredity; sociology studies the real state of the community. Of course, these directions contradicted the ideas of the socialist state and the role of education in it.

However, in the USSR between 1960—1970s, the philological semiotics school played a huge role. The idea of semiotics had been created by humanistic knowledge's desire to be more objective and to converge with the other sciences. In terms of establishing interdisciplinary links, it is noteworthy that the leader of the Soviet algebraic school, *Andrey Kolmogorov*, also dealt with issues of poetry. His enlightenment activity is evident in the fact that he had edited the leading physics and mathematics journal for schoolchildren, named *Quant*, the circulation of which reached 350,000 copies. In addition to promoting entertaining mathematics in the school environment, the journal fulfilled the task of selecting talented students. For this purpose, a task book was included in the journal, and those who succeeded, had the opportunity to participate in the full-time round of the Olympiad in mathematics, which offered the possibility of entering university.

For the Soviet society of the second half of the 20th century, the voice of philologists possessed the authority of sacred truth, neither politicized nor opportunistic. Studies of the Soviet semiotics school's classic scientist *Yury Lotman* on Russian literature, about Pushkin, the Decembrists, and aristocratic culture were perceived as more significant than the editorials of the central communist newspaper *Pravda (The Truth)*.[14] Public lectures

[14] As one anecdote says, there was no truth in the Soviet newspaper *Pravda (The Truth)*, and there was no news in the Soviet newspaper *Izvestia (The News)*. However, this is not strictly true for all the time periods of the USSR's existence.

on history by *Nathan Eidelman* attracted a large audience. Noteworthy was the draft of the Soviet special-scientific and popular literature covering the Decembrist movement (at that historical event, on December 14, 1825, the elite of society, the nobles, rioted against the Tsar at the Senate Square in St. Petersburg). The point is not only that this historical topic was totally accepted by authorities, but, inevitably doomed to failure, the Decembrist riot was almost a form of protection for Soviet dissidents. Thus, one can assume that there was manifested a latent moment of resistance to the regime.

It is generally accepted that humanities scholars tend to popularize to a lesser extent than representatives of the natural sciences (Brockman 1996). However, the popularization activities of Soviet humanists were in demand by the Soviet public. In their voices could be heard the truth about the world as a whole, which was not broadcast by other public communicants. Summarizing the constructive experience of popularizing humanistic knowledge, we can formulate several universal positions. From our point of view, the model of humanistic popularization includes: 1) the breadth of scientific interests of the scientist, ideally—interdisciplinarity; 2) the development of humanistic enlightenment is closely connected with the development of the human sciences themselves, with breakthrough scientific directions; 3) an element of play is necessary, something that was characteristic of the Nietzschean "jolly science".

So, as we noted, 1960—1970 became the best time for popular science in Russia. This was influenced by comprehensive state support and a unique public interest in science. In addition to the propagation of scientific achievements, the popular science journalism of 1960—1970 solved the most important task of combining natural and humanistic knowledge. The former was more actively supported by the state, which as a totalitarian regime was the decisive factor; the latter enjoyed greater support from the liberal intelligentsia.

The printing press system grew both qualitatively and quantitatively. Popular science and art magazines included: *Znanie—sila (Knowledge—Power)* (the title of the magazine is quoting Francis Bacon's aphorism *Knowledge itself is power*), *Himiya i zhizn' (Chemistry and Life)*, *Tehniku—molod'ozhi (Technique to the Youth)*, and others. In these magazines, in addition to articles and essays on various branches of knowledge, works of science fiction, including foreign texts, were published. Revived in early 1960, the popular science magazine *Nauka i zhizn' (Science and Life)*

At least *The Truth* was a good heading for science; the newspaper consistently contained an excellent science department.

became the same sign of a "thaw" and liberalization, as the newspaper *Izvestia (The News)*, or the magazine *Novyiy mir (New World)*. Speaking strictly, the positive changes in the magazine were achieved thanks to *Rada Adzhubey*, who was the daughter of the General Secretary of the Communist Party of the USSR, Nikita Khrushchev, and worked all her life for this magazine. The magazine had a truly unique target audience: it was read by the scientific intelligentsia in order to understand related disciplines. The magazine maintained a strong interaction with its readers, and circulation at its height reached up to 3 million copies a month. By comparison, the circulation of *Scientific American* in the 1980s had already reached 7 million copies a year, which is more than 580 thousand per month. Since 1970, on the basis of the Smithsonian Institute in Washington, the *Smithsonian* popular science magazine began publication, which, due to its variety of materials, was closest to the universal type of the Russian-Soviet popular science magazine. Thanks to a unique magazine model that combines an interdisciplinary approach with themes and material objects that are the subject of study, and exhibits of the Smithsonian Institution and its museums, the *Smithsonian* is considered one of the most popular media in the U.S.

Only at the end of 1960s, television began to play a key role in the Soviet media outlets. Popular science has gained mass popularity thanks to the TV program *Ochevidnoe—neveroyatnoe (The Obvious—The Unbelievable)*. This was aired from 1973 until the death of the presenter, Sergey Kapitsa, in 2012. Sergey Kapitsa was the program's presenter, with few interruptions, for almost 40 years. He was the son of the above-mentioned Nobel Prize winner for physics Peter Kapitsa, who after the 1917 revolution lived in Cambridge, where Sergey was born. English science had always remained exemplary for them both. Sergey Kapitsa followed his father's scientific specialization but then decided to enter the field of popular science. All started with the fact that Sergey Kapitsa compiled a preface to scientific works—the anthology of the "quintessence of knowledge". He was not admitted to the Academy of Sciences, as television activity was not considered entirely worthy of a true scientist. However, thanks to Sergey Kapitsa, the Soviet audience, not only became acquainted with advanced Soviet science through the exposition of leading scientists, but were also plunged into an atmosphere of genuine debate, i.e. there was a polemic in this program not found anywhere else on Soviet television at the time. Like his father, who connected different people in informal communication, Sergey Kapitsa became a media personality, connecting different branches of knowledge, as well as science and art. He said that one should strive to "overcome professional selfishness".

The format of the broadcast changed: first, it was in the genre of television lectures, but this monologue format did not take root on Soviet television. In its classic version, it was a one-hour weekly program in the form of a conversation with an invited guest. Significant is the release of the program in 1977, constructed as a visit by Sergey Kapitsa and the famous Soviet ballerina Maya Plisetskaya to the halls of the Museum of Fine Arts in Moscow. Stopping in front of one of the largest collections of Impressionist paintings in the world, Kapitsa and Plisetskaya discussed the turn from the classical to the neoclassical picture of the world that occurred in the late 19th century. In this respect, the theory of relativity and avant-garde art turn out to be—and this is what the Soviet spectator should understand— phenomena of the same order.

Soviet television had an educational base. This was facilitated by the circumstance that airtime had no monetary expression for television production; there was no concept of media business or private property. The main oral channel for the dissemination of enlightenment became the monopolistic Society, named: *Znanie (Knowledge)*. The Society began its activities in 1947 and was active until the collapse of the USSR. The main form of activity for the *Knowledge Society* was public lectures, and the lecturers were given by the best Soviet scientists. The Society also published popular science literature.

The 1960s were marked by a certain romanticism in Soviet life; it was a time of public hopes for building socialism with a "human face". It is really noteworthy that Khrushchev's "thaw" began with popular science literature: *Iskateli (The Seekers)* a novel by *Daniil Granin*, had been written after Stalin's death. Already in 1956, there was a movie made of this novel. At the center of the novel is a young scientist who, for the sake of truthfulness to scientific truth, was not afraid to challenge the bureaucratic system of organized science. A similar plot, related to the opposition of the scientist and the social environment, became very popular at that time (e.g. Anatoly Agranovsky's publicism, Vasily Aksyonov's prose). The issues around the social responsibility of scientists became, in the second half of the 20th century, among the most important in many leading countries (as, for instance, in the novel by the, very popular in the USSR, writer Kurt Vonnegut—*Cat's Cradle*). In 1965, a cult Soviet film appeared on the screens, also based on Granin's novel *Idu na grozu (I'm on a storm)*, about scientists. The profession of the scientist was directly correlated with public service. Thus, literature and cinema are important channels for the mediatization of science; in this direction, they were able to express a truly significant social problem. During the "thaw" in the USSR, various subcultures, which were actually destroyed in the years of Stalin's terror, began to take shape again. In numerous expeditions (geological, folklore, archaeological) a friendly and professional informal environment was restored: the heroism and romanticism of knowledge had been formed.

The main scientific and technological innovation, exciting in 1960 Soviet society, was space. Soviet boys wanted to become nothing else but cosmonauts. Scientists were obliged to popularize science (this was part of their social work), but then profile journalists appeared. The most famous space journalist was *Yaroslav Golovanov*. In addition, he wrote several popular science books. For example, *Etudes About Scientists* combined essays on world-renowned scientists, with essays on outstanding Russian scientists (Archimedes, Galilee, Copernicus, Lomonosov, Lobachevsky, Pavlov, Sechenov, Tsiolkovsky, Edison, etc.). Science and science popularization, even in a confrontational world, created the prerequisites for processes of integration and the formation of the open world idea. Golovanov begins the author's preface to the book with a quotation containing the statement of the Soviet physicist Sergey Vavilov:

> "The history of science cannot be limited by the development of ideas—it should equally apply to living people, with their characteristics, talents, dependence on social conditions, country, and era. In the development of culture, individual people have and continue to maintain incomparably greater significance than in the general socio-economic and political history of mankind...Therefore, it's clear that the life and work of advanced people is a very important factor in the development of science, and their biography is a necessary part of the history of science ..." <...>
>
> But Pythagoras is not a drawing, Newton is not a formula, Pavlov is not a brilliant experience. These are people, these are fates, and these are the characters. I thought that besides Kepler's three laws, I want to know more about whether this person was happy or unhappy. And how tall was Lomonosov? Was Leibniz happy or gloomy? What color were Galileo's eyes? How did Kurchatov smile? <...> But I'm sure that it's harder to love a drawing, formula or law than to love a person (Golovanov 1976, pp. 5, 6).

This example clearly reveals one of the most important features of Soviet popularization: the focus on a man, a scientist, an organizer of science, a teacher. First of all, the personality of the scientist was mediatized, and, through him, the branch of science and even science as a whole. It was according to this principle that the significant movie *9 dnej odnogo goda (9 Days of One Year)* (1962) was filmed. The plot of Michael Romm's movie was devoted to the nuclear physicists' work and was partly based on real events. At the center of the movie are the moral choices of the heroes, their ability to endanger their own lives for the sake of science.

Interest in science fiction continued to grow. In this genre, not only writers continued to work, but, what was very typical for the Soviet system of science communication, so did the scientists themselves. So, in the preface of 1955 to his science fiction novels, the recognized geologist and geographer *Vladimir Obruchev* wrote:

> Everyone is interested in getting at least a general idea of the forms and conditions of past life. This task I also tried to solve in the book I wrote in the form of a science fiction novel. <...> I know only two novels in which such an attempt is made. One is the novel by Jules Verne "Journey to the Center of the Earth" <...> Geological errors in this novel prompted me in 1915 to compose "Plutonium". Prior to this case, I have not written anything for young readers yet and I did not intend to do it. The second is the novel by Conan Doyle, in which travelers across South America discover a high, very difficult plateau, cut off from the surrounding countryside and inhabited by primitive people <...> A good science fiction novel must be believable, should inspire the reader with the conviction that all described events can, under certain conditions, take place, that in them there is nothing supernatural, miraculous. If in the novel piles of different miracles take place it is no longer a novel, but a fairy tale for kids (Obruchev 1986, pp. 6–7).

But if in the first third of the 20th century the science fiction element was closely connected with the artistic-fictional element, then in the era of space exploration social problems were actively realizing themselves in science fiction. In this sense, the creative path of the famous Soviet fiction writers, the brothers *Arkady and Boris Strugatsky*, was indicative. From the original utopian works, full of faith in scientific progress and scientific fellowship: *Ponedel'nik nachinaetsya v subbotu (Monday Begins on Saturday)*, they passed to anti-utopias: *Ulitka na sklone (Snail on the Slope)*, *Trudno byt' bogom (It's Hard to Be God)*. The pathos of the Strugatskys' mature and later works was no longer technocratic, but purely humanistic. A person does not really need the icy cold of the vast cosmic spaces, which, by mastering, a person more and more transforms and destroys (both the surrounding nature and his own personality). This intention was clearly expressed in the fantastic story *Piknik na obochine (Picnic on the Sidelines)*. In 1979, the no less famous Soviet film director *Andrey Tarkovsky* made the movie *Stalker* based on this story. Earlier, in 1972, Tarkovsky made *Solaris* based on the novel of the same name by the Polish science fiction writer and futurist *Stanisław Lem*. In these Tarkovsky's movies, a certain revolution in the understanding of the future of mankind was well expressed. Comprehension of the secrets of nature must be accompanied by moral development; otherwise, the process of scientific knowledge can lead to disaster. These issues are extremely important today. The prospect of artificial intelligence, discoveries in the field of the human genome and bioengineering—these are global challenges of the present, which science fiction writers have predicted long ago. In the satirical story of the Soviet science fiction writer Ilya Varshavsky, under the name of "Roby", the colleagues from the laboratory of radio electronics give to the protagonist on his anniversary a robot to help in the household. As a result, the robot turns out to be a real monster, because it transfers purely mathematical formulas to everyday life. Thus, the genre of science fiction has huge prognostic potential.

In 1960s, science entered the Soviet people's mass consciousness. The songs of the songwriter and bard Vladimir Vysotsky sarcastically reflected exciting popular science TV programs. Later, already during "perestroika" (the second half of 1980), similar processes began to take place on Soviet television. For example, one of the educational programs, "The Fifth Wheel", released on Leningrad TV, was called "Lenin—Mushroom". In it, the presenters parodied serious popular science programs; they argued that Lenin was nothing more than a narcotic fungus and a radio wave. In terms of Soviet citizens' confidence in the media, it is interesting to note that not all viewers realized that this was a hoax. The collection of short stories and novellas of the above-mentioned Ilya Varshavsky, "Molecular Café" (1964) opens humorously with: "Edification for Science Fiction Writers of All Times and Peoples, From Beginners to Veterans Inclusive". The tradition of such humorous recommendations is well known: these are parodic letter books, dream books, etc. When Stanisław Lem arrived in Leningrad in the early 1960s, he was given the opportunity to read a folder of Varshavsky's unpublished stories. According to the legend, Lem replied, "I never thought that all of Western fiction would fit into one folder".

In 1961, one of the most popular movies, *Amphibian Man*, appeared on the screens in the USSR. It was based on the novel by the above-mentioned science fiction writer Alexander Belyaev. The idea of the film is that it is possible to create an underwater republic on the seabed, where the idea of social justice would be realized. The first inhabitant of the ideal underwater world is Ichthyander, who is able to breathe underwater, thanks to a medical experiment. The social meaning of the movie realizes itself through the plot of romantic love, but the heroes are separated not just by the density of water, but by social obstacles. The very action of the movie takes place in an exotic setting: in a stylized Buenos Aires. To participate in the filmmaking, it was planned to invite the famous French oceanographer and film director Jacques-Yves Cousteau, popular in the USSR. The movie turned out to be very spectacular, with a fascinating plot. In 2018, the Oscar Award was received by the movie *The Shape of Water*. It presents variation on an archetypal love story between different anthropological creatures, one of which lives in water. In the Soviet version, the scientist makes a breakthrough discovery—creates a person who can breathe underwater. According to the plot, villainous capitalists want to take advantage of the results of this discovery for criminal commercial purposes. It must be borne in mind that the USSR did not fight against the West as such, against the people; it fought against the capitalist regime. The inversion of history is manifested in the fact that, in full conformity with the burrows of Soviet propaganda, in the Guillermo del Toro movie, scholars themselves become enterprising capitalists, far from moral norms. Returning to positive knowledge, we note that the depths of the world ocean are much less studied today than the secrets of the cosmos. In this sense, we still do not know much about our own world or the place of the person within it.

During the period of 1980—1990s, the situation in the country was changing radically. Western brands (*National Geographic, Geo, New Scientist*, and others) began to occupy vacant niches in the new and open Russian media market. The wave of their success was due to the fact that Russian people began to discover the open world, and consumer culture began to form rapidly. These magazines in Russia were more like a segment of the lifestyle media. At the same time, classical Soviet popular science magazines were reduced to eking out a miserable existence. That unique audience of lovers of knowledge was lost. Sincere interest in science was a characteristic and highly positive feature of the Soviet regime, with all of its socio-political and economic problems and shortcomings. For domestic science and the system of its dissemination, extremely difficult times arrived at the end of the 20th century.

> But the fascination with the scientific religion was not in vain. Too festive was the spirit of free creative work. The temples of science with their exhibitions of nonconformists, with the songs of bearded bards, with a previously unpleasantly cheerful habit, turned into museums. But it was in them grew a slightly self-confident, ironic elite.
>
> The scientific religion lost its adherents; society was disappointed in another myth. But the privileged learned class remained. It remains to cherish your privilege: remember that twice two=four (Vail and Genis 2018, pp. 125–126).

And a couple of words about the elite: various Soviet scientific research institutes accumulated a large number of scientific functionaries, who were not seriously engaged in science, but they succeeded in creating a specific Soviet system for the organization of science. Such scientists, who had a great entrepreneurial potential, went into business during "perestroika" (reconstruction). The most striking example is Boris Berezovsky (who died in 2013, under unclear circumstances, in London). In Soviet times, he worked at the Institute of Management Problems, in the Academy of Sciences of the USSR. His scientific career was based on the establishment, on the one hand, of the internal communications within the scientific community; and on the other hand, the external communication with the nomenclature that was engaged in production. Subsequently, in 1989, on this basis, Berezovsky's company "Logovaz" emerged as a joint Soviet-Swiss enterprise, laying the foundations of his financial power. When "perestroika" began and was required to really determine the state of society and the ways of its development, it turned out that only scientists were able to give at least some intelligible forecast. In the second half of 1980, there was such a unique media phenomenon as "doctoral journalism". The socio-political press began to print the thoughts of many representatives of science (economists, historians, philologists, etc.), who discussed the transition to a market economy, as well as the difficult history of a country in the 20th century.

As a result, in the heyday of popular science, 1960—1970s, the word of a scientist, the words about a scientist and his/her work had a special value and status. Popular knowledge was broadcast on various media channels (from purely official to liberal youth literature). Science was surrounded by a halo of creativity and freedom. Science was something more than just science; it fulfilled the function of religion in a secular society. However, such a belief in science, technological development and overall progress has, as sociologists believe, a downside. Such a cult is associated with the underdevelopment of the institutions of society. Nevertheless, popular science not only determined the high intellectual level of the Soviet population, it also contributed to the de-ideologization of the consciousness of the Soviet intelligentsia.

References

Agar, John. 2012. *Science in the 20th Century and Beyond.* Cambridge: Polity Press.

Brockman, John. 1996. *The Third Culture: Beyond the Scientific Revolution.* New York: Simon & Schuster.

Golovanov, Ya.K. 1976. *Etyudy ob uchyonyh* [Etudes about Scientists]. Moscow: Molodaya gvardiya.

Obruchev, V.A. 1986. *Plutoniya. Zemlya Sannikova* [Plutonium. Sannikov Land]. Chelyabinsk: Yuzhno-Ural'skoe knizhnoe izdatel'stvo.

Vail, Peter, and Genis, Alexander. 2018. *60-e. Mir sovetskogo cheloveka* [60s. The World of the Soviet Man]. Mocsow: AST.

PART III

POSTGRADUATE PROGRAMS IN SCIENCE COMMUNICATION, AND OPEN SCIENCE

In place of the fashionable ideas of the "end of history" and the "end of science" in the recent past, the widespread idea of "the death of universities" has emerged. Its supporters insist that, rather than training the intellectual elite, universities are becoming purely profit-oriented commercial enterprises. Meanwhile, science journalists and communicators are trained at universities, and this specialty is considered elitist throughout the world.

Training science journalists is an objective need for the state, society, science and the media itself: the Russian media market demands science journalists. However, until recent times in Russia, science journalists were virtually untrained. In a situation of internal competition between Master's programs within a single university space, it is necessary to form a meta-program, principally based on an interdisciplinary approach. The new "Popular Science Journalism" Master's program has become such a program and is interdisciplinary at both the conceptual level and in terms of its implementation. It was designed by the author of this book and has been implemented since 2014 at the School of Journalism and Mass Communications at Saint Petersburg State University. Its concept is determined by the requirements of the moment: a pressing social demand to improve the scientific picture of the world in the public mind and audience education. In the West, scientific journalists are trained at the best universities. American and British science journalists constitute the elite of the journalistic profession, precisely because they are able to understand and objectively present complex issues. In this case, these kinds of Master's programs should be specialized and at the same time—within the selected profile—interdisciplinary in nature.

In the West, those who want to change their specialism often apply for a Master's Degree. In Russia, the emphasis is shifting from Bachelor's to Master's education. In the United States, as a universally acknowledged leader in the modern world in many aspects of establishing communication between science and society, considerable attention is paid to the training of science communicators. Relevant programs exist in the best universities, mainly within Schools of Journalism and their Master's programs. Such programs operate at Columbia University, Stanford University, California University at Berkeley, Michigan State University, etc. However, the foundations of Russian and American Master's programs design are based on different principles. Russian education is traditionally based on the canon; meaning that in the Russian educational environment, as a starting point, the cultural and historical principles are significant. Existing programs in American universities are designed to fulfill the needs and requirements of the specific territory and state in which the university is located. The wide external environment of the specialism obtained is introduced into the educational context of American education. In this way, pragmatism and specialization, peculiar to the whole system of American life, are manifested. In contrast to the American experience, the Russian educational system continues to maintain the fundamentality of preparation. The external environment is not so actively involved in the learning process, largely remaining in the field of the students' separate work or additional activities.

Western scientists are generally renowned for their ability to create and strengthen information events that determine the presence of scientific problems within a public field. Here's how one Russian science journalist writes about this:

> In the Western culture, American and British, the element of communication between scientists and society has become traditional. What does this mean? Every scientist knows that his research depends on funding, including that from the local budget. Therefore, if any professor, as I always tell students, at any university in the United States, conducts an experiment and sees that the frog was jerked at the foot and squeezed 8 times, and in Russia—only 5 times, he immediately organizes a press conference, invites the Governor, tells him about the study and adds, "Mr. Governor, it was thanks to you that we managed to reach such incredible heights in this matter". The press will write about the frog and the scientist. The Governor will cry from pride and happiness. For the next year, he will increase the funding of this laboratory, because the press conference convinced him: the money was not spent in vain (Strel'nikova 2003).

However, for Russian journalism education, Western European analogs are more important than American ones (Morrison 1997).

In Russia, the process of scientific knowledge promotion has been associated with the development of audience consciousness, based on knowledge as such. Intrinsically, popularization techniques and methods are largely invariant. Their mastery should be superimposed on some scientific base; and from these premises, students may receive a quite different basic education: philological, medical, physical and so on. For the "Popular Science Journalism" Master's program, natural scientists and experts in the humanities, physics, and poetry are equally important, something that follows from its conception; and this program, first of all, prepares universal journalists, only then concentrating on specialization in specific scientific areas.

Thus, the purpose of the new Master's program "Popular Science Journalism" is to educate universal science journalists with a general scientific theoretical base and the practical skills to create scientific media texts for different types of media.

The objectives of the Master's program are: (a) to generate a comprehensive knowledge of science theory and history in terms of scientific convergence; (b) to create objectified ideas in the graduates concerning the main channels and mechanisms of science popularization from the standpoint of both history and modernity, based on the experience of both Russian and the foreign media; (c) to generate the graduates' practical skills in creating scientific and educational media texts.

Promotion of the project started a few years before its realization. This occurred on several fronts. Firstly, it was publicly introduced in the form of press conferences ("Science-lunches"). Secondly, publications in the federal and regional Russian media were prepared. Thirdly, a professional orientation of the students was conducted (as a part of Open Days and a special course). Fourthly, establishing and maintaining an interaction with other departments within the university (Department of Physics).

The training of students faces a number of serious problems. The main problem is of a universal scientific nature: Different fields of knowledge are highly fragmented among themselves. There is a deficit of fundamental scientific training in university programs. In the humanities' educational cycle such reductionism is due to two main factors: (1) the rapid development of non-classical and applied humanities majors (such as tourism, advertising, and others), (2) the overall reduction of hours devoted to humanistic disciplines of the federal component of the curriculum.

If we talk about journalism education, the professional community is increasingly leaning toward the position that the strengthening of scientific training for future journalists is actually necessary. In this way, it is imperative to take into account one of the main features of Russian media education in contradistinction to the West. In Western universities and schools of journalism, the major portion of teaching hours and classes are allocated to the formation of students' skills to work with information and make reports for different types of media. Whereas in training future journalists in Russia, impacting on their ability to form an opinion is dominant. In other words, in accordance with the well-known Russian expression: if the "poet in Russia is more than a poet", the journalist is not only an organizer of communication.

The purpose of the Master's program "Popular Science Journalism" is to promote a broad academic approach, thus, the developed program is not only interdisciplinary but also interdepartmental. The profiles of existing Master's programs are taken as a point of reference here rather than the department and faculty. The main thrust of the program is humanistic.

The primary areas of the "Popular Science Journalism" Master's program are: (a) general scientific, (b) history, theory and practice of scientific popularization, (c) the poetics of scientific and educational media texts. The main courses on the curriculum are as follows: "Introduction to Methodology and History of Science", "Modern Natural Science", "History of Popular Science Journalism", "Contemporary Scientific and Educational Film: Types and Genres", "Popularization of Science in Print Media", "Popularization of Science in Audio-Visual Mass Media", "Travelogue Discourse", "Creative Studios", "Environmental Journalism": the last course is especially important. Similar Master's degree programs in Western Schools of Journalism often have a specialization in environmental journalism. Overcoming the technogenic understanding of civilization in favor of that which is nature-friendly marks the modern philosophical understanding of the world:

> Even this imperative presupposes that we decide that humanity and nature deserve to survive. The task remains to provide arguments in favor of this decision, to have this decision develop into a common conviction in open discussion, and eventually to consolidate it over generations. The new impression possesses four advantages over the Kantian one. First, it makes what was taken for granted by Kant - the existence of humanity and nature—an object of duty; second, it makes us responsible for what concerns us and which on the other hand is barred by the Kantian imperative; third, it corrects the anthropocentric narrowness of Kant's definition of the class of possible addressees of human duty. Fourth, and finally, it reminds us of our finiteness, in contrast to a utopian tendency peculiar to a technical-science civilization; it protects us from an illusionary understanding of human freedom and thus converts us to the possibility of using our true freedom (Sitter 1988, p. 217).

So, what are the learning objectives within the framework of the Master's program?

Task 1. Graduates should be capable of a general scientific level, which itself serves as a natural barrier to pseudoscientific knowledge. Basic disciplines—"Introduction to the Methodology and History of Science"—open the Master's degree program curriculum. It is aimed at establishing a common understanding of the scientific process in its unity: diachronic and theoretical. The idea of science convergence, scientific revolutions, with regard to science ethics, the humanistic inspection of the field of high technologies has been established as a methodological criterion. The following major historical periods of the scientific picture of world formation are distinguished: Antiquity, the Renaissance, classical science and the techniques of modern times (18—19th centuries), non-classical science (end of the 19th century—the first half of the 20th century).

The "History of Popular Science Journalism" module promotes an additional and in-depth understanding of the given coordinate system. Within the framework of science popularization history, it is considered in conjunction with the development of science and the socio-cultural situation. This module is designed to reveal the historical and typological models of domestic popular science journalism, which can be applied in the present. Within the course, the "History of Popular Science Journalism", students complete the following task: they must create a popular science journal. More precisely, they are expected to form a specific journal model, its content and design. Their projects are then posted on the faculty website. In addition, within this historical discipline, students write a paper: "Course Research Paper in the History of Popular Science Journalism: The 18th—20th centuries". The idea of the convergence of the humanistic and natural sciences is integral to the various disciplines of the curriculum.

Task 2. The program is called "Popular Science Journalism", not just "Science Journalism or Communication". It is designed to neutralize the possible disparity between different areas of knowledge. The program emphasizes the adaptation of scientific information for a mass audience through the media, movies, literature and, more specifically through, for example, travelogues. This feature allows the program to stand out from the alternative Master's Degrees available. All disciplines are focused on the practical application of knowledge. "Creative Studios" are totally practice-oriented; they are divided into two modules: science popularization in print media, and audio-visual mass media. Within the first module, students write popular science articles and reviews for the corporative popular science journal "Saint Petersburg University". In the second they prepare the script and an application for a non-fiction television program.

During the final semester, students take "Poetics and stylistics of the popular science text", designed to summarize the theoretical and practical skills already obtained. In this context, the main components of the poetics and style of popular science texts are considered. In the aspect of poetics, these are the basic journalistic genres of the popular scientific text: news, popular scientific articles, interviews, reports. Also architectonic and composition, headline, and lead, the pyramidal structure of the popular science media text are studied. In the aspect of language, a major focus of interest is the technology of adapting scientific statements to a mass audience; the ways of working with conceptual apparatuses, sources and references; correct citation; editorial work on the text. All of these techniques are considered both in terms of media criticism-critical analysis of journalistic texts and, at the same time, the creation of the students' own media texts.

Within the framework of purely practical disciplines, the cycle accentuates the following aspects of the scientific and educational media text: its code multiplicity, the intention of hobbies and entertainment at the same time, the interaction of verbal and nonverbal components. One of the most important skills, process formation which is given special attention, is the ability to increase occasions where information can be widely disseminated. The key problem of science coverage lies in the fact that the agenda created by the scientific community requires serious processing in order to become an effective media message.

Task 3. Unfortunately, educational and research institutions lack the tools to struggle against pseudo-science. One mission for the graduates on this Master's program is to carry out this kind of activity in terms of specific journalistic actions (by means of investigative journalism, the formation of adequate public opinion, etc.). Countering pseudo-science in the humanities is no less important than in the natural sciences. Different fields of knowledge converge at this point but differ in their constitutive role in terms of the formation of ideas about the past and the future. The compensatory function of scientific popularization also has considerable potential in that it allows the expression of that which is difficult to discuss within the rigors of pure scientific discourse.

The scientific and popular press is an intellectual resource, which should satisfy the need for knowledge throughout the world. There is a hypothesis regarding a half-life of knowledge—the time period after completion of training, during which half of the original assimilation of information is lost.

The decline in the prestige of science, its chronic under-funding (especially in the humanities), the disunity of the scientific community and its poor mediatization—all these factors contribute to a crisis situation in which Russian science finds itself now. Scandals involving science create the typical information of the science media landscape. The media gladly tell stories about counterfeit dissertations, an endless series of sensational scientific discoveries, finally, practical advice a la: "British scientists have found out". Daily academic life: conferences, expeditions, defense of dissertations, as is commonly believed, do not create information occasions and mostly remain outside media coverage. The ability to such information occasions, bringing it closer to the needs of the audience, requires professional skill.

In the mid-eighties, in Russia, the cultural and commodity exchange with the West amplified and expanded the ideology of an affluent society. With the interests of a consumer society in general, and in Russia in particular, a focus on glamour and tourism are now established; while popular science journalism generates appropriate values for its audience. Worthy science popularization should be able to strike the right balance between entertainment and the cognitive aspects. The primary audience of the popular scientific press remains quite elitist.

The graduates of the Master's program should have some kind of universal cognitive base. In addition, they should be able to generate scientific information occasions, in order to bring them closer to the audience. Meanwhile, the controversy in the public sphere regarding science popularization is centered on those issues that have been long known. The issues around whether journalists could or should popularize science are already actively discussed in public, and a positive answer to this question is already accepted. In the Soviet era, the objective, which was to raise the level of the mass audience to a scientific one, science was successfully popularized by the scientists themselves as well as by journalists. A similar situation exists in the Western press, and, thus, the Master's program is developing in this direction.

Graduates should, then, be able to act in two main directions. Firstly, they should be able to mediatize science itself. In this case, their focus is on the advancement of scientific knowledge, revealing an approach to building modern scholarly communication. In the second case, they are expected to think about the audience and the development of its image of the world. In Russia, in the past, scientific education always solved these two problems.

In our view, a modular approach is the best principle for the Master's program's adequate formation within the framework of its structural association (faculty, institute). And in this case, the best students are actually selected. At the end of the first year, the number of students reduced by approximately one third. This happened despite the fact that these students at admission had the highest score. The reason for this situation is the difficulty of the program. According to an opinion poll among students, they consider that this Master's program profile develops both cognitive skills and practical experience. In the context of other journalistic Master's programs at Saint Petersburg State University, this program is close academically to a number of programs: "Documentary Film: Creativity and Technology", "Historical Journalism", and (to a lesser extent) "Journalism and the Culture of Society". These programs have a lot in common in terms of curriculum. It makes sense to combine them into a single modular unit with a variable component. Such modularity would provide horizontal mobility within the programs.

It is well known that science can only progress within the context of scientific schools and a viable scientific environment. Scientific popularization is the same, but it develops within the different context of promotional formats and integrated marketing communications. Therefore, the considered Master's program works closely with the science journalists' club: "Science Matrix". It has been working in St. Petersburg for more than ten years and organizes press conferences with leading scientists.

The main intended outcome of this program is as follows: graduates of the Master's program will certainly be in demand. We already have an employment offer from the press service of academic institutions and popular science journals. It is important that, as a result of students' research, and journalistic and teaching practice, they will be offered jobs appropriate for the level of a Master's degree. Graduates of the Master's program need to fill the vacant segment of science journalists. It is very important that they work not in journalism in general, but more specifically in science journalism. Graduates of Russian journalism faculties generally have fewer difficulties finding a job in their specialty than their Western contemporaries (see Lukina and Vartanova 2017).

The science journalists' and communicators' training should ultimately develop the competencies necessary for the modern professional capable of dealing with the idea of the world as infinitely diverse and variable.

> Conclusions in science are strange in that they are not forever. "Nothing is static, nothing is final, everything is held provisionally," said Jocelyn Bell Burnell, the astronomer who discovered pulsars. Put plainly, no sooner do you get used to one theory than somebody comes up with a better one (Webb 2013, p. 198).

In the process of training science journalists and communicators, one of the universal problems of education is reduced: the necessity to orientate towards the average rather than gifted students. Let us repeat that science journalists are precisely the elite. Applicants to such programs, as a rule, have a good background. However, in the Russian reality, the future of such exclusive programs is problematic. It is planned that applicants will enter expanded generic profiles, for example, "Journalism", within which secondary specialization will be carried out.

The science journalist's primary mission is to discover science for society, and then to make it public. This task is the first priority in our modern global world. One recent action that received a public response was a public statement by The University of California (UC). The University published a Recommendation "Call to Action", which listed the main tasks necessary to transform science communication, urging colleagues to solve the problem jointly. The University of California believes that it is time to work out a collective strategy to reduce university spending on scientific literature while supporting the movement toward open access for scientific information. As noted at the university press release, many institutions and university consortia in Europe believe that the problem of costly subscriptions to scientific journals can be solved by switching to immediate open access. In Europe, there are several such projects: the *OA2020 Initiative*, the activities of the German *Projekt DEAL*, and the *No Deal No Review* movement in Finland. All these initiatives are aimed at redistributing library budgets, moving away from paying subscriptions to scientific articles, and toward publications in the public domain:

> We believe the time has come to address these issues head-on through a combined strategy that places the need to reduce the University's expenditures for academic journal subscriptions in the service of the larger goal of transforming journal publishing to open access. Through our renewal negotiations with publishers, we will pursue this goal along two complementary paths: by reducing our subscription expenditure and investing in open access support. Such an undertaking will be neither easy nor without short-term pain for the University, its faculty and students. However, even if we were able to maintain all of our current subscriptions, the long-term prospects for beneficial change outweigh the merits of attempting to cling to an untenable and undesirable status quo (Negotiating Journal Agreements at UC 2018).

Modernity is marked by a variety of online science forms, which nevertheless have a national specificity. Russian scientists, on the one hand, have access to global databases; on the other hand, in conditions of confrontation with the West, these opportunities are being reduced. At present, the problem for researchers searching for full scientific texts is quite acute. Subscription to leading scientific journals is expensive and, unfortunately, even the leading Russian universities do not always have them. There are other problems. Since 2017, the resource that allowed the download of the full texts of articles, *Sci-Hub*, has blocked access to Russian users.

At the same time, leading Russian universities are doing a lot of work aimed at creating an "open university". The National Electronic Information Consortium, in partnership with a number of universities, is currently implementing the project: *The national aggregator of open repositories of Russian universities*, supported by the Presidential Foundation of the Russian Federation. The project assumes the creation of a unified platform for open access repositories, accumulating the scientific works of Russian scientists.

The leading universities in the country are increasing the number of different digital collections, thematic portals, and subject indexes. Internal databases of all publications by faculty members are created. In parallel, centralized resources of online courses have appeared. However, these positive changes mainly relate to the initiatives of the universities' senior management. In the conditions of a certain disintegration of the different areas of Russian science, and the overall difficult situation, there is a search for new ways for cooperation, initiated by the scientific community itself.

One such way is the outlet to the public sphere. Starting in 2009, we see in Russia the steady growth of various popular science and educational resources (network portals, video films, cultural and educational sites). There is also a rise in natural science blogs and sites. Representatives of humanistic knowledge also seek to unite both in institutional and non-institutional groups. The new media, combining journalistic, enlightening, and educational functions; to which can be added the function of storing information (depository), has developed. These hybrid media represent research data and, at the same time, open-up science for different target groups. They make science fashionable, both intellectually and emotionally attractive. This is what will build the "one world" of the future.

To date, a highly constructive situation has developed in the Russian public space in terms of access to dissertations. Mostly thanks to the efforts of official authorities, the procedure for defending dissertations has become public for the first time after a long period. At the same time, increasing publicity has contributed to the consolidation of civil society and representatives of the scientific community efforts. Accordingly, open access resources containing information about the dissertation thesis and its author are affiliated both with official, commercial and public organizations.

At the same time, the system of training scientific staff and awarding academic Degrees in Russia is still poorly integrated with the world community. The problem is that this system is hybrid, combining Soviet and modern Western properties. From the Soviet era, a complex two-tier system of scientific Degrees has existed. In the USSR, science was respected, partly in the ancient, magical sense, on one hand, by the state, and, on the other hand, by society. Obtaining a scientific Degree opened many social possibilities. The Russian Degree of "Candidate of Sciences" corresponds to the generally accepted Ph.D. This degree is considered as the third level of a multi-level system of higher education and is awarded as a result of mastering the doctoral educational program accepted in the West, which in Russia is called the "graduate school". In some European countries, there is awarded the highest scientific Degree, named "doctor habilitatus" (Dr. habil. = habituated doctor). It gives the right to occupy a professorship at the university and corresponds to the Russian Degree of the "Doctor of Science". In order to count on a scientific career in Russia, it's necessary to defend not one, but two dissertations.

In different countries, the state has different powers to regulate the procedure for awarding academic Degrees and accredited doctoral (Ph.D.) programs. In such European and non-European countries, like France, Norway, Spain, Sweden, the United Kingdom, New Zealand, state (and authorized by the state) structures play a crucial role in this process. In the USA and Canada, in many ways existing in a single space, this role belongs to the socio-professional organizations. Finally, in the Netherlands, the Academy of Sciences has the right to make the decision. The right to assign academic Degrees is different too. In Russia, and post-Soviet countries, this right is reserved for the state; in the rest of the world – for the universities themselves. Accordingly, in different countries, the information regarding a dissertation's defense and public access to it is formed in various ways. In Russia, the main role in this process belongs to the state.

There are a number of the obligatory stages of the dissertation's publicity before a defense.

1) Publication on the website of the organization, where the work was prepared. Therein is published the full text of the dissertation, the author's abstract of the dissertation, information about the official

opponents, and the leading organization, preparing a review for the dissertation, as well as the place and time of the dissertation's defense. From 2012, at Saint Petersburg State University all dissertation defenses are accompanied by an online translation on the university website (Ph.D. SPbSU 2017).

2) Publication on the website of the High Attestation Commission under the Ministry of Education, and Science of the Russian Federation. Since 2006, the High Attestation Commission publishes on its website the same information, only in a reduced form, and with reference to the organization's website (the High Attestation Commission 2017)
3) The main results of the dissertation research should be published in the form of scientific papers in the list of peer-reviewed scientific journals. The High Attestation Commission revises this list on a regular basis.

And there are a number of the obligatory stages of the dissertation's publicity after defense.

1) Obligatory dispatch of the dissertation abstract to the leading libraries of Russia, and also the National Library of Belarus (National Library of Belarus 2017).
2) The full written text of the dissertation is sent to the Russian State Library. The main library of the country provides public access to the digitized abstracts, and the theses themselves. Site navigation is available in English (Russian State Library 2017).
3) A mandatory copy of both the thesis and the author's abstract is sent by regular mail to the Russian Book Chamber (Russian Book Chamber, 2017). In accordance with law, a certain number of copies of any printed publication (books, brochures, periodicals, dissertations, musical notes, geographic maps, and atlases, etc.) must be sent to the Russian Book Chamber.

Other (additional) online channels for the dissertation's distribution.

1) Official state resources. The most important is the Scientific Electronic Library (eLIBRARY.RU). Materials placed in this electronic library are included in the Russian Scientific Citation Index. This Index is a National bibliographic database of scientific citations, collecting more than 12 million publications of Russian scientists, as well as information on citing these publications from

more than 6,000 Russian journals. Another resource, having a printed version, is the network encyclopedia "Famous Scientists" (Famous Scientists 2017). This provides the biographical data of scientists and is a project of the Russian Academy of Natural History. It also publishes the main results of each dissertation's research.

2) Commercial resources. To such belongs the "Library of Dissertations" (DsLib.net). It provides both free and paid access to the thesis' text and publishes information about upcoming dissertation defenses. At the same time, monetary deductions to authors of dissertations are declared, but, in reality, they do not work. We can give another example. "Cyber Leninka" (CyberLeninka 2017) positions itself as a scientific open access electronic library, the main tasks of which are the popularization of science and scientific activity, public control of the quality of scientific publications, the development of interdisciplinary research, the modern institute of scientific review, and the increasing citation of Russian scientists. "Ciber Leninka" is built on the basis of the open science paradigm.

   The fundamental problem is that theses and abstracts are both published "on the rights of the manuscript", in other words, they don't formally have a copyright. Consequently, private distributors do not, in fact, pay any royalties, which violates the norms of morality, if not the law.

3) Public organizations resources. The main resource for exposing plagiarized or falsified dissertations is the network "Dissernet". It positions itself as "a free network of experts, researchers, and reporters, who dedicate their work to exposing scammers, falsifiers, and liars. The participants of the community cooperate based on the principles of a network distribution of labor and the use of modern computer technologies, to counteract illegal frauds and infringements in the field of scientific and educational research, especially in the process of defending dissertations and appropriating academic degrees in Russia" (Dissernet 2017). More than 10,000 dissertations have been analyzed by "Dissernet" since 2013. Based on "Dissernet" requests, the High Attestation Commission has made numerous decisions to deprive candidates of their academic degrees.

One of the latest scandals was the case, initiated by "Dissernet", of stripping the Minister of Culture of the Russian Federation, Vladimir Medinsky, of his academic title of Doctor of Historical

Sciences. His dissertation was devoted to the problem of objectivity in the foreign researchers' coverage of Russian history. The main argument of the expert community was that Medinsky has considered the methodology of the historical research as corresponding to the national interests of the state. With this approach, historical science turns into a servant of the dominant ideology, as happened in Soviet times. This approach is typical of historical pseudoscience. Nevertheless, at the final meeting of this case, which took place on October 20, 2017, the Presidium of the High Attestation Commission ruled to allow Minister V. Medinsky to retain his Doctoral Degree. Such disclosures of the political establishment's representatives often occur in Germany (thanks to "VroniPlag Wiki", as well): this is an indication of a healthy society. In Russia, this case additionally indicates that scientific reputation itself becomes irrelevant, which is a huge problem.

So, as a partial conclusion: the most positive effect of the current situation is that never before has science in Russia been so public. This result was achieved due to the efforts both of the state (its scientific policy) and society. The next step will be the integration of this openness into the Western system. But, at the present time, the procedure for awarding academic degrees is beginning to change. A number of leading Russian universities and scientific organizations receive the right to self-award scientific degrees. Thus, at St. Petersburg State University, at present, defenses are already being passed under the new rules. On the one hand, in this way Russian universities become more integrated into the Western system. The supervising role of the state is canceled. The dissertational councils do not act on a permanent basis but create in accordance with the subject matter of the concrete work. The defenses could take place in English, and – most importantly – the academic degree awards directly. But such innovations have caused some criticism. It is believed that in a democratic society decentralization is extremely constructive. In the scientific policy of modern Russia, this kind of decentralization causes great concern about maintaining a high scientific level. Actually, there is a return to the closed Soviet system of leading universities. One of the most serious concerns: will the thesis not be less public, and the thesis defense process more closed?

References

CyberLeninka. Accessed October 2017. https://cyberleninka.ru/

Dissernet. Accessed October 2017. https://www.dissernet.org/about/

Famous Scientists. Accessed October 2017. https://www.famous-scientists.ru/about/

High Attestation Commission. Accessed October 2017. http://vak.ed.gov.ru.

Library of Dissertations. Accessed October 2017. http://www.dslib.net/

Lukina, Maria, and Vartanova, Elena. 2017. Journalism Education in Russia: How the Academy and Media Collide, Cooperate, and Coexist. In: *Global Journalism in the 21st Century Challenges & Innovations*. Ed. by Robyn S. Goodman & Elanie Steyn. University of Texas, Austin: Knight Center for Journalism in the Americas, School of Journalism, pp. 155–174.

Morrison, John. 1997. The Changing Model of Russian Media and Journalism Education. In: *Journalism & Mass Communication Educator*, 52, pp. 26–34.

National Library of Belarus. Accessed October 2017. http://www.nlb.by/portal/page/portal/index.

Negotiating Journal Agreements at UC: A Call to Action. Accessed July 2018. https://libraries.universityofcalifornia.edu/groups/files/slasiac/docs/NegotiatingJournalAgreementsAtUC_ACallToAction_final.pdf

Ph. D. SPbSU. Accessed October 2017. https://disser.spbu.ru/ph-d-spbsu.html.

Russian Book Chamber. Accessed October 2017. http://www.bookchamber.ru.

Russian State Library. Accessed October 2017. http://www.rsl.ru.

Scientific Electronic Library. Accessed October 2017. https://elibrary.ru/defaultx.asp.

Sitter, Beat. 1988. The New Categorical Imperative: The Ethical Principle for a Technological Age. In: *Revolutions in Science: There Meaning and Relevance.* Ed. by William R. Shea. Sagamore Beach, MA: Science History Publications, pp. 205–220.

Strel'nikova, L.N. 2003. O tendenciyah v mirovoj zhurnalistike i meste nauki v SMI [On the trends in world journalism and the place of science in the media]. In: *Stenogramma master-klassa "Nauka v SMI segodnya. Opyt rossijskoj i britanskoj zhurnalistiki"* [The shorthand record of the master class "Science in the media today. The experience of Russian and British journalism"]. Saint Petersburg [*without* formal editing].

Webb, Jeremy (ed.). *Nothing. From absolute zero to cosmic oblivion—amazing insights into nothingness* / New Scientist. London: Profile Books, 2013.

CONCLUSION

To date, there are no fundamental studies of the history of science mediatization, and not only in the Russian but also in the English-speaking scientific community. Attention is drawn to the fact that Russia is also completely outside of the sphere of attention when researchers investigate science communication across the globe. In this sense, Russia, with its rich history and great potential, is paradoxical outside of the western public field. Russia is only actively represented in the West within the framework of confrontational political models. This study is the first research, both in Russia and abroad, to explore the Russian science mediatization tradition in an international context.

The development of new science and modern forms of science communication began in Western Europe in the 17th century. In relation to the European model of development, Russia moved more slowly and lagged behind, but from about the middle of the 19th century, it became one of the world's leading scientific powers. Historically, the main difference of the Russian popular science system consisted of the primary role of the state, as well as the great importance of the humanistic component. The Western system of knowledge mediatization is much more consistent, commercialized and pragmatic; it is more focused on the promotion of science itself. Traditionally, there were two main channels of science mediatization: oral and written. Each specific historical stage is characterized by these channels' special relationship. For effective communication between science and society, the coordinated work of the main socio-cultural institutions of society: the system of science, education, and enlightenment—is necessary.

From the point of view of the dynamics of development, the popular science press is a unique segment of the Russian press, because historically it suffered less significant changes. The Soviet model of the popular science press was essentially borrowed from pre-revolutionary times; Soviet science in many ways continued the Russian Empire scientific project. In contrast to the socio-political periodicals, as well as other specialized publication subgroups on popular science, pre-revolutionary popular science periodicals continued their lives after the Bolshevik revolution. In this sense the system and typological features of popular science publications in the late 19th—early 20th centuries should be considered classical, being exemplary for the following stages.

From the point of view of scientific achievements, the system of science organization and dissemination, and education in science communication, Russia is an integral part of the Western world. Moreover, in the 20th century Russia managed to achieve an unusually high profile for science in society, which was accompanied by an original development in the forms of science mediatization. The level of popular science development in the USSR was unusually high, indeed, unique. Science replaced religion for the Soviet people. In turn, the Soviet system of science popularization, based on the desire to raise the level of the Soviet people to the scientific, provided an extremely high status and prestige for Soviet science in public opinion, something that largely determined USSR power. Channels of knowledge mediatization included various institutional and non-institutional associations, the press, and fiction. By the end of the existence of the USSR, at that time the most well-read society in the world, every 20th published book belonged to the category of popular science. Since "perestroika" the Soviet popular science press has begun to function again in the general context of its Western analog, thereby largely losing its own identity.

From the historical process standpoint, signs of the popular science revival in modern Russia should lead to the construction of an environment that does not duplicate the Soviet or Western experience, but combines them. The matrix for such an interaction is revealed in university education—science journalists training in the framework of Master's programs. Indeed, there are problems in modern Russian science. One of the symptoms is a failure to decrease scientific emigration from the country. However, despite all the difficulties, Russian science develops and has considerable potential. And one last thing: at all times and within the different national schools, a great scholar forms both in accordance with and also contrary to the system that influenced his becoming. Finally, there is one paradox. In Russia, science popularization cannot be unpopular; that follows from tradition and the peculiarities of the national identity.

APPENDIX

GLOBAL WARMING: PRO AND CONTRA (INTERPRETATION IN THE RUSSIAN PUBLIC SPHERE)

Many global topics are covered in the Russian media space, but not quite as in the West. One of them is the concept of climate change, in relation to which the scientific community still does not have a single point of view. For illustration, we will present the point of view of the well-known Russian independent journalist Yuliya (Julia) Lanynina, Ph.D. in Philology, cooperating with the media, which are considered to be the independent: the radio station *Ekho Moskvy (Echo of Moscow)* and *Novaya Gazeta (Newspaper)*. At the present time, due to threats against her, Latynina is forced to live outside Russia. She is known as one of the brightest critics of the theory of global warming. Julia Latynina expressed her position on the theory of global warming repeatedly and continues to develop and clarify the main points. Let us give as an example a fragment of her weekly author's program *Kod dostupa (Access Code)* on the radio *Echo of Moscow* dated January 30, 2010, dedicated to this issue (https://echo.msk.ru/programs/code/652349-echo/):

"A lot is being written about my position on global warming. A typical text, for example, looks like this: we do not know much about CO2 (carbon dioxide), but you will agree that if we fight against global warming, the environment will be cleaner. But the answer is not true. Why? Because pollution is the real problem. The environment is polluted: the rubbish dump near your home, water—you cannot drink, air—you cannot breathe.

The snag is that any contamination is local. Water can be contaminated with alkalis, acids, dioxins, and mercury, whatever. There is one substance that never pollutes the water. This is carbon dioxide, CO2. You drink it with soda. When you drink any soda, CO2 is dissolved in it, precisely because it is safe, precisely because it is part of the biological turnover of substances, because plants inhale CO2, we exhale it. It is always present in the atmosphere. A tray with dry ice, ice cream is kept in dry ice for the same reason that we do not faint from our own breathing—CO2 is absolutely safe.

The same is true of the air. Here is the city of Norilsk. Two million tons of sulfur dioxide annually, 44,000 tons of lead. What does the city of Bratsk pump into the air? Perfluorocarbons. The city of Dzerzhinsk pumps out almost all the combat chemicals. The city of Astrakhan emits hydrogen sulphide. Air pollution is always local. There is only one gas that all these plants throw out and that does not bother anyone. It's called, oddly enough, CO2.

And when you are shown a tube and told that it's CO2 and it causes pollution, then you have been told a lie twice. First, as I said, CO2 is a gas present everywhere, a gas that is absolutely not poisonous and safe for humans because it is built into the biological turnover of substances. It is invisible, colorless and odorless. Thus, when you see a tube, it's definitely not CO2. Secondly, when a tube is shown to you... let's say if these are the same sulfur aerosols, then, excuse me, they, according to physics, cool the air.

If we want to fight not against pollution, but against global warming, then we need to smoke more, we need to throw out more sulfur dioxide. In the city of Norilsk, sulfuric acid eats everything physically, <…> here they pump out two million tons, but it is necessary to encourage them, to make it so that they can throw out ten million. That's not accidental, because any pollution, as I said, is local. And it is necessary to fight against pollution at the local level. But people who want to fight against CO2, they do not want to go to Norilsk, they want to go to Copenhagen. Therefore, they want to fight against something global.

And the problem is that it is impossible to fight with something global, and the chemical compound that is everywhere cannot be poisonous, it is embedded in the biosphere of the Earth. We cannot smell carbon dioxide, because it does not smell at all. Probably, if life was arranged differently and we, let's say, breathed fluorine, and carbon dioxide was poisonous for us, maybe we would smell it. But since it is not a marker, we cannot smell it. And this is a very important point.

Because when we are told—let's struggle with environmental pollution, and then they say that CO2 is a pollutant, then, in addition to the fact that people are obviously not very good at biology, this is a substitution of concepts. And the substitution of concepts always exists when there is a totalitarian ideology. A totalitarian ideology is when one takes an existing problem—for example, the absence of an 8-hour day for workers—and says, "We will now fight for the dictatorship of the proletariat". No, it is not necessary for the dictatorship of the proletariat; let us simply support the workers.

In fact, climate change and its connection with human civilization is such an amazing thing that I will try to talk about it more and more. The very first thing we see is that the Earth has existed for 5 billion years, during which time the glacial periods were periodically replaced by interglacial periods.

Think of the previous interglacial period, the Eemian period, it started in 130 thousand years BC and continued for 15 thousand years. The world was much warmer than the current one. At the North Pole in summer the water was melting, Scandinavia was an island, in the Thames were hippopotamuses. Apparently, the climate in Siberia was 3–9 degrees warmer than it is currently.

Then the glaciation began, then, as we all know from the history books, 10,000 years ago it was gone. But, in fact, the glaciers began to retreat earlier. However, 12,900 years ago—this is a very accurate dating, a very significant event—it happened in the period of the Younger Dryas—a sudden cold fell upon the Earth.

Why do I concentrate on this specific event? There is in the history of mankind a most important event that occurred—when people moved from gathering to farming. There was not only wheat or rice, which they began to cultivate, but also the notion of private property emerged, because mushrooms in the forest are not private property, while wheat in your area becomes private property.

And a fantastic circumstance is that, as now scientists believe, it was in the Younger Dryas that it happened, during a sudden cold snap. Moreover, it happened with three, and most likely—with four completely different civilizations. This happened in the region of the Middle East, where wheat began to be cultivated, in Central Mexico it was corn, in the middle reaches of the Yangtze—rice, and it seems that even in Ecuador, which no longer distinguished itself in the history of mankind, they cultivated a pumpkin at this time.

These four completely independent cultures, with a completely different future, suddenly took over and moved quite sharply from gathering to farming. It was a global process. And the only thing that was global at that time was a sudden cold snap. They did not have mobile phones; they did not call each other.

By the way, this sudden cooling caused the death of the so-called Clovis culture in America, which failed to adapt, and did not find anything to cultivate. Then we have the climatic optimum Holocene from 9 to 5 thousand years, which again is interrupted by an incomprehensible, very short event around 6200 BC.

This is followed by periods of warming and cooling. Then one of the most famous episodes begins somewhere around 60 BC and lasts almost 600 years. Note that it strangely coincides with the collapse of the two great empires—the Roman and the Chinese, which the Germans and Huns conquer. Then a thousand years ago the medieval climatic optimum begins. At this time in England they were growing grapes; at this time Erik the Red, discovered Greenland and called it "green." The population of Europe at that moment was the same as in the 19th century.

And in the 14th century, a small ice age begins. 1315—the first total hunger all over Europe. This famine, which lasts for several years, and the "black death" epidemic in 1348 halve the population of Europe to such an extent that now in France there are places where fewer people live than in the 14th century. In addition to this terrible extermination of the population, there was a spiritual catastrophe—religious fanaticism began, witch-hunts began, and the mass persecution of Jews began.

By 1600 it is warmer. And from 1645 to 1715 again a small ice age. Europe freezes, on the ice of the Thames they arrange fairs. From 1790 to 1830—a new period of cold weather, by the way, this also includes the army of Napoleon, which came to Russia. For this period, there is also the terrible year of 1816, which is called the year without a summer. This is the year when the frosts killed the May harvest across Europe, when 100,000 Irish died, when snow fell in Taiwan.

At the same time, because of a poor harvest, American culture began to move to the West. The movement to the West began as a response to the year without a summer. And then the short glacial period ends really only in the middle of the 19th century. The temperature begins to rise. It continues until the 1940s. After that, it falls for several decades, and to the extent that television in the 70's, which already exists and already teaches about typhoons, about hurricanes, explains this as a new global cooling. Then, from the mid-70s, the temperature is on the mend, suddenly it turns out that we have global warming.

Why do I tell all this? Because we see as a result that the climate has no norm. There is a shell of the Eemian period, which will say, "So, guys, we now have a sea 6 meters below the norm". The only norm of nature is change.

The second thing is very important: a biological catastrophe for humanity is cooling. Try to explain to the hundred thousand Irish who died after 1816, that they were saved from the horrors of global warming.

The third and even more important thing is that it is not so much the physical change in nature, but humanity's social response to this that is the problem. Here the Clovis culture in the Younger Dryas took and died. And in the Middle East, they learned how to grow wheat. After the famine in 1315, hunting began after all. And after the year without a summer in 1816, the movement of American civilization to the West began. By the way, do you know where the temperature did not change? In Africa; and humanity did not evolve there.

It is quite obvious that everything I'm talking about cannot be explained by anthropogenic carbon dioxide. The medieval climatic optimum was quite accurately conditioned not by the fact that the Vikings in Finland built secret factories and pumped CO2 into the atmosphere.

And I have a strong suspicion that 90% of people who, with placards in their hands, demand a stop to global warming will not be able to answer the question of why the climate is changing. I will try to explain very briefly a few things, why it is most likely changing.

Firstly, we know exactly why the glaciation is taking place. The answer to this question was given by the great Serbian scholar Milutin Milanković (Milankovich). He died in 1958. According to Milankovich, the periods of glaciation or interglacial depend on the amount of heat received by the Earth, especially its Northern Hemisphere, where the mass of land is large. This depends, in turn, on periodic changes of the Earth's orbit associated with precession, with nutation (this is the angle of inclination of the Earth's axis to the plane of the orbit) and with a change in the eccentricity of the orbit. All processes have different cycles: 25 thousand years, 41 thousand years and 93 thousand years. When all this unhappily coincides, a glaciation is obtained. But Milankovich cycles are cycles of global processes. And what about the little stories, such as the small ice age, which had its peaks, which had its ups?

The peak of the small ice age: 1645—1715. You will laugh, to astronomers this date is very well known, it is called the Maunder Minimum. The Maunder Minimum is the time when astronomers did not observe spots on the Sun. They at this time just noticed them; they began to watch very carefully. And absolutely wonderful astronomers— e.g. Cassini— looked at this matter.

Moreover, there was another minimum, the minimum of Dalton, from 1790 to 1830, the very one that the year without the sun fell on, the very one where Napoleon froze in Russia. By the way, both Dalton and Maunder are not self-names, the name given to these periods was given by the great American astrophysicist John Eddie, also in a brilliant work which was published in 1976.

It can only be stated that up to the end of the 20th century there has been an absolute coincidence of the solar activity graphs and these minimal climate changes. For example, solar activity increased from the middle of the 19th century until 1940, when the temperature rose. From the 40's to the mid-70's, solar activity fell, and with it the temperature dropped. This solar activity explains what happens to the temperature in these short periods, while the accumulation of CO2 in the air does not explain this in any way. Because, according to the current hypothesis of climatic alarmists, carbon dioxide affects the climate almost from the middle of the 19th century.

Forgive me, it turns out that a person has not yet invented the car, and carbon dioxide has already influenced the climate. And then the year 1940 began, then the Second World War began, in which all belligerents launched thousands of bombers into the air and built factories and smoked, regardless of any costs. And then the person for some reason ceased to exert influence on the climate and did not render either in the 50's or 60's, when the industrialization of the land was monstrously growing, and the temperature for some reason was falling. <…>

Now the obvious question. Well, it's all fine. In the past, the Earth's climate depended on Milankovich cycles, on solar activity, but it could also change as a result of a volcanic eruption, from a meteorite fall, from the collapse of a glacier into the ocean. Is it that the huge amount of CO2 pumped out by mankind into the air is not a new factor that did not exist before? I will answer this question in the next broadcast. And the answer to this question will also be the answer to the question about the shortcomings of democracy.

But I must speak very briefly about two problems that arise in this case. The first problem is very simple. It is difficult to imagine a demonstration with the words "Down with Milankovich cycles", while demonstrations with the words "Down with global warming" are very easy to imagine.

The second problem is even more terrible. The fact is that the theory of global warming means power, influence, and money. And the money here is even secondary, the main thing is power. A climatologist who studies the cycles of Milankovitch, this is just a tabletop. A climatologist, who undertakes to regulate carbon dioxide, undertakes to command the economy of the whole world. He gets what Genghis Khan, Hitler, and Stalin dreamed of. Agreed, this is not a very pure scientific experiment, when from the answer to the supposedly scientific question depends who you are, you are nobody or the master of the world".

Let us consider the above position through the prism of two polarized estimates. In the first case, Julia Latynina's conclusions are supported; in the second, they are criticized. The first is represented by Ilya Treiger, not an academic scientist, but a full member of the Russian Geographical Society. He is the editor of the information-analytical network edition of *UNIPRESS*, one of the most popular Russian-language publications, published in the United States. His response appeared shortly after the cited broadcast (http://www.russiandenver.50megs.com/latynina.html):

"The essence of the error, in fact, in Julia Latynina's conclusions about the harmlessness of carbon dioxide. <…>

The maximum permissible norm of carbon dioxide content in indoor air is 0.1-0.15%. Studies conducted in the UK in 2007, found that at a carbon dioxide level of 0.1% (i.e., twice as much as normal atmospheric levels), employees feel a headache, fatigue, and cannot concentrate. All this ultimately leads to an increase in the number of hospital sheets and the inability to work productively. That is, even at these concentrations, carbon dioxide becomes a poison, but not yet fatal. <…>

In the infamous Nazi warheads, vans, where the exhaust pipe was directed into the van, people were killed precisely with carbon dioxide, and not with carbon monoxide, as is often believed in everyday life.

Why is carbon dioxide safe in soda or in the form of dry ice?

In carbonated water carbon dioxide enters the human body through the stomach, and not through the lungs. This is the first reason for the safety of such consumption of carbon dioxide. As an analog here you can consume snake venom or a poison such as curare. Both of these poisons can be simply drunk without any danger to health. But the problem arises if these poisons get directly into the bloodstream through the surface of a wound. Similarly, carbon dioxide is one thing when it gets into the stomach, and it's quite another matter when it is breathed through the lungs. <…>

Yes, Julia Latynina made an error. However, this inaccuracy in no way contradicts the journalist's conclusions on both global warming and air pollution. <…>

As mentioned above, increasing the concentration of carbon dioxide in the ambient air doubles the first signs of poisoning. A rise in this concentration to 26 times, as in a diving bell, is already unambiguously deadly. Nevertheless, in this case, no significant increase in temperature is observed either in stuffy rooms or in diving bells, if, of course, we do not consider a certain increase in temperature due to the heat released by people. That is, even a 26-fold increase in the concentration of carbon dioxide does not cause a noticeable increase in the temperature of the surrounding air. And in today's Earth atmosphere, the concentration of carbon dioxide is only 0, 038%, and this is along with all the production emissions around the planet. How can we say with such figures that carbon dioxide in the Earth's atmosphere causes global warming, even if it is a greenhouse gas? In this situation, the statement about the causal relationship between anthropogenic carbon dioxide and global warming requires very serious evidence. And such evidence, even if precisely calculated, has not yet been presented to anyone. Instead of evidence, we have mostly statements that are close in content to political slogans.

So, to refute the arguments of Julia Latynina, just latching onto her inaccuracies concerning carbon dioxide, will not succeed. Those who are interested in global warming will have to refute arguments such as Milankovich's cycles, the minimum of Maunder, the minimum of Dalton, and so on from the set given by Latynina. Only on these grounds can you refute this position, but not on the basis of the fact that the journalist was mistaken about the toxicity of carbon dioxide..."

Let us dwell on one of the alternative positions expressed recently. The expert assessment of a professional climatologist on Julia Latynina's article on climate change in *Novaya Gazeta* (https://www.novayagazeta.ru/articles/2017/06/03/72682-yuliya-latynina-nas-vozmut-teplenkimi) was published 10/07/2018 in the online edition of *Gazeta.ru* (https://www.gazeta.ru/science/2017/07/10_a_10777604.shtml\).

The scientist's report was preceded by the next preamble, setting out the negative point of view on the Latynina position: "In what ways Julia Latynina was wrong, speaking of a worldwide conspiracy of climate scientists, explains one of the leading authors of the IPCC assessment reports, Sergey Gulev, Professor, Corresponding Member of the Russian Academy of Sciences, Head of the Laboratory of Ocean and Atmosphere Cooperation and Climate Monitoring of the Institute of Oceanology".

Sergey Gulev begins with the following clarification: "The request to comment on Julia Latynina's article on the climate in *Novaya Gazeta* caught me off guard because I never tried to comment on non-professional opinions about the physical mechanisms of climate and weather formation. Reading the article itself made a depressing impression of illiteracy and unfounded arguments, as well as self-confidence and frenzy. The article by Latynina is very poorly structured, the author jumps from one argument to another, often repeats, so it is really difficult to comment. However, I'll try".

Here are some of Gulev's arguments:

"*Thesis 1:* There is a doctrine of global warming, the author calls the theory. This doctrine (or theory) is developed by the IPCC, established in 1988.

It is nonsense. There is no doctrine of global warming; there is a field of physics called climate physics. The subjects of its study are the observed climatic changes at different scales, the physical description of their mechanisms and the modeling of these changes with the aim of a possible forecast.

Diagnosis of the observed changes began in the late 19th century, thanks to Vladimir Keppen; then in the 1930s Guy Stewart Callendar related the observed changes to the concentrations of CO2. The physical description of the response of temperature to various factors is presented in the late 1960s independently by Budyko and Sellers. The simple model proposed by them was later used to analyze the growth of surface temperature discovered in the 1970s (and published in leading scientific journals).

In response to this call the IPCC was created, before it there were no questions, it had only "terms of reference". <…>

Thesis 5: 'As warming increases, the number of natural disasters will increase. So this is lies'.

It's not a lie. Physically, this is due to the nonlinear response of the degree of stochastization of the system, even to small changes in internal energy. This is confirmed by both the processing of observational data and climatic models. The number of articles is in the thousands. The intensity of extreme precipitation in the mid-latitudes is increasing, although the long-term amounts of precipitation change rather weakly. <…>

Here again, I quote Latynina, "The only thing that can result in warming in high latitudes is to reduce the number of hurricanes because hurricanes in the most general case arise because of the difference in temperature of air masses at the equator and in temperate latitudes".

Absolute nonsense. Hurricanes do not arise because of the difference in air temperatures at the equator and in temperate latitudes. They arise as a result of a diabatic signal in the tropics, and with warming; the effect of "expanded tropics" will lead to an increase in the number of intense tropical cyclones. Because of the same effect, the influence of Hadley's circulation on the trajectory will increase, and they will be more directed to the north. <…>

Thesis 6: (expressed by Latynina earlier). Modern climate models cannot predict it, because they cannot even predict the weather for three to five days in advance.

This is a complete misunderstanding again. Indeed, climate models for the structure of equations are quite similar to the weather forecast models. However, there are differences in predictability for short periods and for very long periods. The model includes (if very simple) seven equations for two environments (ocean and atmosphere) plus equations for ice, processes on land and some other mechanisms. <…>

Thesis 7: (expressed implicitly here, but earlier it was repeatedly expressed explicitly). Warming is more beneficial than harmful, better for people, agriculture, etc. So do not fight it.

I do not presume to judge here, not being a specialist in adaptation to climate change (apparently, Latynina is). Physicists give detailed information about what changes can occur, where and why. Other experts are exploring how ecosystems will change under the influence of climate change. Finally, the third (economists, sociologists) are studying the possibilities of adapting society and the economy to such changes.

As a layman, I can say that adaptation is probably a more correct strategy than trying to reduce the effect of climate change. In this sense, my attitude, for example, to the *Kyoto Protocol*, and *The Paris Agreement* is more skeptical. There are many uncertainties in the very scenarios of anthropogenic emissions (Retrospective Concentration Pathways), which is connected with the uncertainty of the forecasts of economic development. If you find 100 scientists who will reasonably explain to you that the climate becomes warmer, only a few (probably less than 20) will agree that it is necessary to act as written, say, in the Paris Agreement on climate.

Finally:

I do not think that in some way I will change the opinion of Ms. Latynina. After reading and listening to some of her materials on other topics, I am sure that she is not one who can change her mind.

I do not think either she or I will want to meet in direct discussion. It would be a conversation between the blind and the deaf".

The scientist's conclusion, obviously, calls into question Julia Latynina's position. However, it is Latynina who, as a journalist, poses a politically incorrect and inconvenient question that is not discussed by the scientific community: is not the "concept of global warming a multi-billion dollar swindle of world bureaucrats"? In this sense, popular science journalism connects with the socio-political, and performs a compensatory function, actualizes those issues that, for various reasons, are not accepted in academic science.

Bibliography

Akopov, A.I. 2012. Popularization of science as a crucial link in the propaganda and advancement of research. In: A.I. Akopov *Journalism: Research Papers, Articles, Notes, and Essays: Collection of Papers*. Rostov-on-Don: North Caucasus Higher Education Research Center of the South Federal University Academy of Pedagogical and Social Sciences, pp. 112–118.

Agar, John. 2012. *Science in the 20th Century and Beyond*. Cambridge: Polity Press.

An Almanac for Moderns. 1938. By Donald Culross Peattie. Including a New Introduction by the Author and Wood Engravings by Asa Cheffets. Printed in Washington for the Members of The Limited Editions Club.

Balashova, Yu.B. 2007. *Shkol'naia zhurnalistika Serebrianogo veka* [School Journalism of the Silver Age]. St. Petersburg: Saint Petersburg University Publishing House.

Balashova, Yu.B. 2017. Debate about Scientific Popularization in Russian Public Sphere (Based on Grey Literature Material). In: *The Grey Journal: An International Journal on Grey Literature,* vol. 18, n. 1, pp. 179–181.

Beliy, Andrey. 1989. *Na rubezhe dvuh stoletiy. Vospominaniya* [At the turn of two centuries. Memories]. Vol. 1. Moscow: Hudozhestvennaya literature.

Berdyaev, Nikolay. 2005. *Filosofiya svobodyi* [Philosophy of freedom]. Moscow: AST.

Berkowitz, Carin, & Lightman, Bernard (eds.). 2017. *Science Museums in Transition: Cultures of Display in 19th-Century Britain and America*. Pittsburgh: University of Pittsburgh Press.

Blum, Deborah, Knudson, Mary, and Henig, Robin Marantz. 2006. *A Field Guide for Science Writers*. New York: Oxford University Press.

Bowater, Laura, and Yeoman, Kay. 2013. *Science Communication: A Practical Guide for Scientists.* Hoboken, NJ: Wiley-Blackwell.

Bradley, Joseph. 2012. *Voluntary Associations in Tsarist Russia: Science, Patriotism and Civil Society*. Moscow: Novyj hronograf.

Brockman, John. 1996. *The Third Culture: Beyond the Scientific Revolution*. New York: Simon & Schuster.

Bucchi, Massimiano. 1998. *Science and the Media: Alternative Routes to Scientific Communications*. Abingdon: Routledge (Taylor & Francis Group).

Bynum, William. 2013. *A Little History of Science*. New Haven and London: Yale University Press.

Capp, Bernard. 1979. *English Almanacs, 1500–1800: Astrology and the Popular Press*. Ithaca, New York: Cornell University Press.

Cooper, Caren. 2016. *Citizen Science: How Ordinary People Are Changing the Face of Discovery*. New York: The Overlook Press.

Curth Hill, Louise. 2005. The Medical Content of English Almanacs: 1640—1700. In: *Journal of the History of Medicine and Allied Sciences*, 60 (3), pp. 255–266.

Dawkins, Richard. 2011. *The Magic of Reality: How We Know What's Really True*. New York; London; Toronto; Sydney; New Delhi: Free Press.

Daum, Andreas W. 2009. Varieties of Popular Science and the Transformations of Public Knowledge: Some Historical Reflections. In: *A Journal of the History of Science Society*, 100 (2), pp. 319–332.

Foucault, Michel. 2004. *The Archeology of Knowledge*. London: Routledge.

Golovanov, Ya.K. 1976. *Etyudy ob uchyonyh* [Etudes about Scientists]. Moscow: Molodaya gvardiya.

Graeme, Donald. 2013. *The Accidental Scientist. The Role of Chance and Luck in Scientific Discovery*. London: Michael O'Mara Books Limited.

Gregory, Jane, and Miller, Steve. 1998. *Science in Public. Communication, Culture, and Credibility*. Cambridge, MA: Basic Books.

Horgan, John. 1996. *The End of Science: Facing the Limits of Knowledge in the Twilight of Scientific Age.* Reading, Mass.: Helix Books / Addison-Wesley.

Izvestiya russkogo obshchestva lyubitelej mirovedeniya [News of the Russian Society of Amateurs of the World]. 1912, 1.

Jaspers, Karl. 2010. *The Origin and Goal of History*. Abington: Routledge.

Lazarevich, E.A. 1981. *Populyarizatsiya nauki v Rossii* [Science popularization in Russia]. Moscow: Moscow University Publishing.

Lukina, Maria, and Vartanova, Elena. 2017. Journalism Education in Russia: How the Academy and Media Collide, Cooperate, and Coexist. In: *Global Journalism in the 21st Century Challenges & Innovations*. Ed. by Robyn S. Goodman & Elanie Steyn. University of Texas, Austin: Knight Center for Journalism in the Americas, School of Journalism, pp. 155–174.

Markov, A.V. 2015. *Rozhdenie slozhnosti. Evolyutsionnaya biologiya segodnya: neozhidannyie otkryitiya i novyie voprosyi* [The birth of complexity. Evolutionary biology today: unexpected discoveries and new questions]. Moscow: Corpus.

Matheson, Donald. 2005. *Media Discourses: Analysing Media Texts*. Maidenhead, Berkshire: Open University Press.

Merton, Robert K. 1973. *The Sociology of Science: Theoretical and Empirical Investigations*. Chicago: University of Chicago Press.

Moles, Abraham. 2008. *Socio-dynamics of culture*. Moscow: Publishing house LCI.

Muravyeva, Irina. 2004. *Vek moderna: panorama stolichnoy zhizni* [The modern age: panorama of life in the capital city]. Saint Petersburg: Pushkin Foundation Publishing House, 2.

Mussell, James. 2007. 19th-Century Popular Science Magazines, Narrative, and the Problem of Historical Materiality. In: *Journalism Studies*, 8 (4). pp. 656–666.

Obruchev, V.A. 1986. *Plutoniya. Zemlya Sannikova* [Plutonium. Sannikov Land]. Chelyabinsk: Yuzhno-Ural'skoe knizhnoe izdatel'stvo.

Russell, Nicholas. 2010. *Communicating Science: Professional, Popular, Literary*. New York: Cambridge University Press.

Schiele, Bernard, Claessens, Michael, & Shi, Shunke (eds.). 2012. *Science Communication in the World: Practices, Theories and Trends*. New York; London: Springer.

Scientific American. 1900, May 19, Volume 82, Issue 20.

Sloan, David Wm., Startt, James D. (eds.). 1996. *The Media in America: a history*. Northport: Vision Press.

Toffler, Alvin. 1990. *The Third Wave*. New York: Bantam Books.

The Magazine of Popular Science, and Journal of the Useful Arts. 1837. Volume the third. London: John W. Parker, West Strand.

Vaganov, Andrey. 2014. *Spiral' zhanra: Ot "narodnoj nauk"' do razvlekatel'nogo biznesa: Istoriya i perspektivy populyarizacii nauki v Rossii* [Spiral of the genre: From "folk science" to entertaining business: History and prospects of popularization of science in Russia]. Moscow: Lenand.

Vail, Peter, and Genis, Alexander. 2018. *60-e. Mir sovetskogo cheloveka* [60th. The World of the Soviet Man]. Mocsow: AST.

Webb, Jeremy (ed.). *Nothing. From absolute zero to cosmic oblivion—amazing insights into nothingness* / New Scientist. London: Profile Books, 2013.

Weber, Max. 1946. Science as a Vocation. In: Max Weber. *Essays in Sociology*. New York: Oxford University Press, pp. 129–156.

INDEX